TUMOUR LOCALIZATION
WITH RADIOACTIVE AGENTS

... International Atomic Energy Agency:

AFGHANISTAN	HOLY SEE	PHILIPPINES
ALBANIA	HUNGARY	POLAND
ALGERIA	ICELAND	PORTUGAL
ARGENTINA	INDIA	QATAR
AUSTRALIA	INDONESIA	REPUBLIC OF SOUTH VIET-NAM
AUSTRIA	IRAN	ROMANIA
BANGLADESH	IRAQ	SAUDI ARABIA
BELGIUM	IRELAND	SENEGAL
BOLIVIA	ISRAEL	SIERRA LEONE
BRAZIL	ITALY	SINGAPORE
BULGARIA	IVORY COAST	SOUTH AFRICA
BURMA	JAMAICA	SPAIN
BYELORUSSIAN SOVIET SOCIALIST REPUBLIC	JAPAN	SRI LANKA
	JORDAN	SUDAN
CANADA	KENYA	SWEDEN
CHILE	KOREA, REPUBLIC OF	SWITZERLAND
COLOMBIA	KUWAIT	SYRIAN ARAB REPUBLIC
COSTA RICA	LEBANON	THAILAND
CUBA	LIBERIA	TUNISIA
CYPRUS	LIBYAN ARAB REPUBLIC	TURKEY
CZECHOSLOVAKIA	LIECHTENSTEIN	UGANDA
DEMOCRATIC KAMPUCHEA	LUXEMBOURG	UKRAINIAN SOVIET SOCIALIST REPUBLIC
DEMOCRATIC PEOPLE'S REPUBLIC OF KOREA	MADAGASCAR	
	MALAYSIA	UNION OF SOVIET SOCIALIST REPUBLICS
DENMARK	MALI	
DOMINICAN REPUBLIC	MAURITIUS	UNITED ARAB EMIRATES
ECUADOR	MEXICO	UNITED KINGDOM OF GREAT BRITAIN AND NORTHERN IRELAND
EGYPT	MONACO	
EL SALVADOR	MONGOLIA	
ETHIOPIA	MOROCCO	UNITED REPUBLIC OF CAMEROON
FINLAND	NETHERLANDS	
FRANCE	NEW ZEALAND	UNITED REPUBLIC OF TANZANIA
GABON	NIGER	
GERMAN DEMOCRATIC REPUBLIC	NIGERIA	UNITED STATES OF AMERICA
GERMANY, FEDERAL REPUBLIC OF	NORWAY	URUGUAY
GHANA	PAKISTAN	VENEZUELA
GREECE	PANAMA	YUGOSLAVIA
GUATEMALA	PARAGUAY	ZAIRE
HAITI	PERU	ZAMBIA

The Agency's Statute was approved on 23 October 1956 by the Conference on the Statute of the IAEA held at United Nations Headquarters, New York; it entered into force on 29 July 1957. The Headquarters of the Agency are situated in Vienna. Its principal objective is "to accelerate and enlarge the contribution of atomic energy to peace, health and prosperity throughout the world".

Printed by the IAEA in Austria
September 1976

PANEL PROCEEDINGS SERIES

TUMOUR LOCALIZATION WITH RADIOACTIVE AGENTS

PROCEEDINGS OF AN ADVISORY GROUP MEETING ON
TUMOUR LOCALIZATION WITH RADIOACTIVE AGENTS
ORGANIZED BY THE
INTERNATIONAL ATOMIC ENERGY AGENCY
AND HELD IN VIENNA, 9-13 DECEMBER 1974

RC 270
A 32
1976

INTERNATIONAL ATOMIC ENERGY AGENCY
VIENNA, 1976

TUMOUR LOCALIZATION WITH RADIOACTIVE AGENTS
IAEA, VIENNA, 1976
STI/PUB/451
ISBN 92–0–111276–9

FOREWORD

Nuclear medicine has been preoccupied with tumour localization from its very beginning. Advances in scintigraphic techniques and new radiopharmaceuticals have increased the relative importance of scintigraphic studies in tumour detection. Interest was significantly heightened in 1968 by observations on the accumulation of radiogallium in tumours and in other tissues, and at about the same time new in-vitro tests indicated the possibility of serological detection of cancer. Finally, advances in oncology have increased the needs for defining tumour extension and for finding suspected tumours.

The present agents for detecting tumours by scintigraphic techniques are not as good as they should be. The ideal agent should be highly specific for cancer with an avid uptake in the tumour, and the fraction which is not taken up by the tumour should be rapidly excreted from the body. The compound should be inexpensive, simple to prepare, heat-stable, easily sterilized, have low chemical toxicity, be reliable and only cause a low radiation exposure. These criteria may never be met at the same time in one radiopharmaceutical.

It will be necessary to gauge the efficacy of the different agents now available or to be prepared in the future. This means that nuclear medicine has to move away from past patterns into a more orderly and well-reasoned stage of clinical trials on new agents. Statistical methods have to be designed to evaluate whether a new agent is 50%, 10%, or perhaps 5% better than another well-established procedure. These problems are of dimensions that make it unrealistic to think that nuclear-medicine specialists can come up with generally accepted guidelines for evaluating new agents and their clinical importance which pay due credit to differences in instrumentation and other aspects of protocol. But, as far as possible, an attempt should be made to achieve a valid comparison and to have similar criteria for the statement that a tumour is present or not.

The real importance of scintigraphic techniques for detecting tumours lies in the impact which a positive or negative scan will have on the cure rate or morbidity of the patient. It may be questioned what changes in life, survival or cure rate have been brought about by brain scintigraphy which, at the present time, is the most valuable of the scintigraphic procedures. Even if brain scintigraphy is of great comfort to the neurologist and to the surgeon and of no discomfort to the patient, the final evaluation must be linked to the clinical consequences of the scintigraphic study.

There is a need for better agents, especially in view of improvements in tumour therapy and with screening procedures in mind, and there is a need for better methods of evaluating the different reagents. The meeting of which this book is a record was convened on 9–13 December 1974 to summarize the results obtained by the use of tumour-localizing agents and to delineate some possible trends in further development.

The 15th General Conference of Weights and Measures has adopted the SI units for **activity**

$$1 \text{ Bq (becquerel) (dimensions: } s^{-1}) = 1 \text{ dis/s}$$
$$\text{i.e. } 37 \text{ G Bq} = 37 \cdot 10^9 \text{ Bq} = 1 \text{ Ci}$$

and for **absorbed dose**

$$1 \text{ Gy (gray)} = 1 \text{ J/kg}$$
$$\text{i.e. } 10 \text{ m Gy} = 0.01 \text{ Gy} = 1 \text{ rad}$$

CONTENTS

Tumour-localizing pharmaceuticals (IAEA-MG-50/10) 1
   V.R. McCready, N.G. Trott
   Discussion 19
Factors affecting uptake of radioactive agents by tumour and other tissues
(IAEA-MG-50/14) 29
   R.L. Hayes
   Discussion 40
Radioiodine-labelled compounds previously or currently used for tumour localization
(IAEA-MG-50/2) 47
   W.H. Beierwaltes
   Discussion 53
Tumour localization using radiomercury-labelled compounds (IAEA-MG-50/12) 57
   C. Raynaud, D. Comar
   Discussion 60
Clinical and experimental studies of selenium-75-labelled compounds: a review
(IAEA-MG-50/21) 63
   A.H.G. Paterson, V.R. McCready
   Discussion 66
Tumour scintigraphy with gallium-67: present status (IAEA-MG-50/9) 69
   H. Langhammer, G. Hör, K. Kempken, H.W. Pabst, P. Heidenreich, H. Kriegel
   Discussion 79
Tumour localization with technetium-99m (IAEA-MG-50/5) 83
   H.J. Glenn, T.P. Haynie, T. Konikowski
   Discussion 91
Tumour localization with radionuclides of indium (IAEA-MG-50/11) 93
   H.J. Glenn, T.P. Haynie, T. Konikowski
   Discussion 99
Labelled antibiotics as tumour-localizing agents (IAEA-MG-50/8) 103
   D.M. Taylor, V.R. McCready
   Discussion 111
Radiolanthanides as tumour-localizing agents (IAEA-MG-50/4) 113
   K. Hisada, A. Ando, Y. Suzuki
   Discussion 122
Tumour localization using compounds labelled with cyclotron-produced short-lived
radionuclides (IAEA-MG-50/13) 125
   D. Comar
   Discussion 131
Summary of General Discussion 135

List of Participants 141

TUMOUR-LOCALIZING PHARMACEUTICALS

V.R. McCREADY, N.G. TROTT
Departments of Nuclear Medicine and Physics,
Royal Marsden Hospital and Institute of Cancer Research,
Sutton, Surrey,
United Kingdom

Abstract

TUMOUR-LOCALIZING PHARMACEUTICALS.
The varied characteristics of malignant and benign tumours make it highly improbable that any single radiopharmaceutical will be able to localize in all lesions. However, despite the disappointments of much work in the past, it is possible that better agents may result from fresh analyses of the requirements of tumour-localizing agents. In this report the chief applications of such agents are reviewed, these including the early detection of small tumours, differential diagnosis, the exact delineation of tumours and the study of tumour metabolism. The physical properties required of the radionuclide used as a tracer are reviewed, and it is noted that as there have been few fundamental changes over several years in the basic design of imaging equipment which is commercially available, conclusions reached ten or so years ago remain valid, although now a greater range of radionuclides with the desirable properties of short half-life, photon energy 100 – 300 keV, and low intensity of charged particle emission are available. The statistics of tumour detection are considered in relation to analyses of images obtained in clinical measurements. Biological aspects of the problem are discussed with attention to morphology, cell kinetics, tissue type and vascularity. Various possible mechanisms of uptake of particular agents in tumours are considered and it is concluded that in our present state of knowledge of the chemistry and physiology specific to malignant tumours one must still rely largely on intuition in choosing agents for study.

The use of radiopharmaceuticals to detect and diagnose tumours is a subject which has recently been the theme of several conferences. Yet a survey of the work presented so far gives little indication that the techniques and radiopharmaceuticals are of undoubted clinical value [1]. One reason for this may be that we are using generalized approaches to problems which demand a more specialized attack. Thus, the purpose of this paper is first to look at the medical problems involved and then to explore the underlying pathology and physiology in an attempt to define the features which might help in designing appropriate radiopharmaceuticals.

An initial problem lies even in the definition of a "tumour". In one sense a tumour is a palpable or visible lump. On the other hand the word "tumour" is often used to indicate malignancy — being a lesion which eventually kills the patient. Willis [2] defines a tumour "as an abnormal mass of tissue, the growth of which exceeds and is unco-ordinated with that of the normal tissues and persists in the same excessive manner after cessation of the stimuli which evoked the change". Benign tumours show no tendency to invade surrounding tissues whereas malignant tumours always invade surrounding tissue and show a progressive growth which results in death in untreated cases. A vast variety of malignant and benign tumours have been described and their differing appearances and behaviours documented. Thus, the concept that a single radiopharmaceutical could be produced which would detect all these lesions is unlikely to be realized. However, it is important to at least try to investigate the problems involved in the hope that future work is less based on guess-work than some of the past attempts.

The characteristics of an ideal tumour-localizing pharmaceutical will involve both physics and chemistry. Firstly, the choice of the radionuclide is somewhat easier to discuss partly because more research has been carried out already and also because the parameters and instruments

TABLE I. CURRENT MEDICAL PROBLEMS IN IMAGING

Organ	Current technique Radionuclide	Main problem	Tumour/lesion Hot or cold	Comments
Brain	Brain imaging/dynamic study $^{99}Tc^m O_4^-$	Differential diagnosis Small and posterior fossa tumours	Hot	EMI scan will help but isotopes of value initially for position and to screen patients
Parathyroid	Static imaging ^{75}Se-methionine	Detection of small tumours	Hot	X-ray techniques are difficult
Thyroid	Static imaging ^{131}I, $^{99}Tc^m$, ^{123}I	Differential diagnosis	Cold	Ultrasound can help
Lung	Perfusion imaging $^{99}Tc^m$-MAA	Differential diagnosis Detection of small lesions	Cold	X-ray resolution good but differential diagnosis poor
Liver	Static/dynamic imaging $^{99}Tc^m$-colloid	Differential diagnosis Detection of small lesions	Cold	Ultrasound can help in both cases but of less help in estimation of liver size
Spleen	Static imaging $^{99}Tc^m$-red cells	Differential diagnosis Detection of small lesions	Cold	Ultrasound can help
Pancreas	Subtraction imaging ^{75}Se-methionine	Differential diagnosis Detection of small lesions	Cold	X-ray techniques are difficult
Skeleton	Static imaging $^{99}Tc^m$-PP	Differential diagnosis	Hot	Better than X-ray
Lymph nodes	Static imaging $^{67}Ga/^{111}In$-bleomycin $^{99}Tc^m$-colloid	Cold area physiology or tumour?	Hot	Difficult X-ray techniques but has good resolution Some areas inaccessible
Kidneys	Static/dynamic imaging ^{197}Hg or $^{99}Tc^m$ compounds	Detection of small lesions Differential diagnosis	Cold	Ultrasound/X-ray techniques better Ultrasound probably best
Occular tumours	^{32}P	Differential diagnosis	Hot	Isotope method poor EMI scan hopeful
Breast	No good isotope methods	Detection of small lesions	Hot	Ultrasound probably best

involved are known. Secondly, the choice of the chemical compound is more difficult to assess, both because relatively little work has been done on this aspect and because the biological requirements are likely to vary with each tumour and organ.

It is worth considering first the various medical and physical problems involved where radioisotope tests are being used currently. This shows the need for a more specific approach in each situation. While, ideally, we always require high activities with a low radiation dose for the best images the biological factors change with each clinical problem (Table I).

In summary there is:

1. A requirement for an early diagnostic agent where very small tumours are being sought, the patient is probably normal and young, and the radiation dose is important.
2. The problem of differential diagnosis where cancer is possible or likely — a lesion having been already discovered by clinical examination, X-rays or radioisotope imaging.
3. The need for exact delineation of the primary tumour and the ability to find secondary deposits once the diagnosis of cancer has been made.
4. A need for radiopharmaceuticals where the concentration in tumours reflects metabolism and therefore can be of help in the assessment of the response to treatment.

It would be impossible to discuss in detail all the problems just outlined. However, some general comments can be made. Firstly, it would be valuable to have a single agent which could detect the spread of any particular type of cancer and find lesions at a stage where treatment would be effective, say, when the lesion is not more than about 2 cm in diameter. It would be useful if this agent were specific, so enabling a differential diagnosis between malignant and benign lesions to be made and, finally, it would be valuable if the concentration of radioactivity indicated the biological activity of the lesion.

Thus, the problems encountered in developing such a radiopharmaceutical involve two main areas — the physics of the detection of small lesions deep within the body with varying background radioactivity and the biological and chemical factors.

1. GENERAL ANALYSIS OF PHYSICS ASPECTS

The choice of the ideal radionuclide for use in tumour localization involves general principles which are common to many of the techniques now widely used for radioisotope imaging. Comprehensive attempts to bring together all the variables in a radionuclide imaging system were initiated some ten years ago particularly by Dewey and Sinclair [3], Beck [4], Brownell [5] and Matthews [6] (Table II). Matthews incorporated, in addition to the other factors such as characteristics of the detector system, the problem of radiation dose to the patient. Beck's detailed analysis can be used as a guide for the design of scanning systems. However, Harper et al. [7] made the most dramatic use of the work by demonstrating that the optimum photon energy for the detection of a one-inch tumour at the centre of the head was between 100 − 200 keV. The rather unexpected conclusion that $^{99}Tc^m$ had especially suitable properties at once stimulated its introduction into the field of radionuclide imaging.

1.1. Imaging equipment

Although ten years have elapsed since the Anger gamma camera was still a comparative novelty, the basic design with its sodium iodide detector has not changed significantly until recently. Similarly, although scintillation scanners now generally have sodium iodide detectors of

TABLE II. VARIABLES AFFECTING CHOICE
OF RADIONUCLIDES FOR TUMOUR
LOCALIZATION

Figure of merit $F = f(A, B, C, D)$

A : Physical properties of radionuclide

B : Biophysical and chemical properties of nuclide affecting distribution and turnover in body

C : Tissue transmission and scattering

D : Characteristics of imaging system

A	Energy and intensity of photons / charged particles Half-life Purity Availability
B	Practicable compounds In-vivo distribution including turnover rates
C	Depth of site of interest Absorption coefficients of local tissues
D	Sensitivity to local and distributed sources Energy discrimination Intrinsic resolution for given photon energy Resolution time Focal distance

127 mm diameter rather than 76 mm the basic design is still similar to that used over ten years ago. Thus, one would expect analyses carried out at that time to be equally valid today (Table III). However, developments which might be expected to modify these conclusions include:

(a) Improved electronics, making possible the use of more precise energy selection and hence more control over the proportion of scattered radiation that is received. This can extend downwards to a limited extent, the usable range of photon energies, and improve the geometric resolution for a given energy. Progressive improvements to the Autofluoroscope should also be noted.

(b) New and modified types of camera [8], including multi-detector, positron, semi-conductor, image intensifier cameras and spark chambers. Interesting work is being carried out on all these devices and some commercial instruments have been produced although no serious competitors to existing machines are widely available. The semi-conductor camera has the intrinsic merit of high energy resolution, so that exclusion of scatter is simplified, and high geometric resolution is possible. Image intensifier and particularly spark chamber systems seem potentially only suitable for use at relatively low photon energies; it has been found possible to use triggering devices with intensifier systems to provide some energy discrimination but the chief applications are likely to be in high speed dynamic imaging rather than tumour localization. Spark chambers have proved of particular interest in work with ^{125}I (35 keV). At this energy one cannot exclude scattered radiation by energy discrimination but self-absorption in the tissues can provide some compensation.

(c) Improved methods of data analysis, including digital computer techniques. Up to the present these seem to have had only a limited value in routine tumour imaging, although they have assisted in defining the performance of imaging systems and have provided more flexible methods of display.

(d) Multi-detector positron-annihilation radiation systems and the new version of the Auto-fluoroscope are also likely to prove of particular value in dynamic studies; however, both instruments are suitable for use at photon energies of about 0.5 MeV and so probably above the optimum level for the Anger camera and may therefore have applications for tumour imaging with certain radionuclides.

(e) Increase in the diameter of sodium iodide detectors in mechanical scanning equipment to 20 cm increases the sensitivity but often at the expense of depth of focus. With such detectors one may be able to use lower activities or shorter scanning times for many studies, but no fundamentally new principle which might affect the choice of radionuclide is involved.

In brief, one is still thinking in terms of the basic Anger camera and rectilinear scanner.

1.2. Choice of radionuclide

In attempting to discuss the choice of radionuclide for use with available imaging systems we may usefully quote Paul Harper [9]: "There exist several hundred radioisotopes to work with and the question as to which one will prove the most useful in a given situation is often a very complex problem; in fact it is so complex that frequently the decisions must be made on the basis of inspiration rather than logic."

Fortunately, in expressing this opinion Paul Harper is in line with some current views on the methodology of science [10], and we need not be too embarrassed by his conclusion.

Figure 1 shows the range of radionuclides commonly used for scanning at the present time, set out to illustrate the types of radiation emitted, the range of photon energies and the half-lives. The popularity of electron capture and isomeric transition radionuclides with photon energies $0.1 - 0.3$ MeV is notable. Table IV illustrates the proportion of the energy emitted as non-penetrating and as penetrating radiation for some of these radionuclides.

Thus, the most useful radionuclides will have the features — moderate photon energy, short half-life and a low proportion of "non-penetrating radiation", i.e. electrons, positrons and X-rays less than 15 keV. Beyond this, however, other factors, such as the complexity of the photon spectrum, technical ease and cost of production, and radionuclide purity are all primarily physical factors which may influence our choice. Finally, the potential of the radionuclide for incorporation in useful chemical forms will determine its particular role, if any, in imaging, even when all physical factors are in its favour.

Thus, from the physical point of view the popularity of $^{99}Tc^m$ is clearly understandable, although for chemical reasons this radionuclide cannot satisfy all our requirements.

Although the general analyses of Beck and others contain all the essential parameters, it may nevertheless be worth while to consider in a little more detail the significance of certain characteristics of radionuclides, and to illustrate the discussion with some practical examples.

1.3. Half-life and dosimetry

The radiation dose received by a particular organ (named as the "target") in a patient is determined by the quantity that can be expressed as $(\tilde{A} \Delta \Psi)/m$ where $\tilde{A}$ is the time integral of activity at a particular site constituting the radiation source, Δ is a constant which depends on the total energy emitted per nuclear transformation of the radionuclide, and Ψ is the fraction of the energy emitted from the source which is absorbed in the target organ of mass m. If drastic simplifying assumptions are made, we can establish an optimum value for the effective and hence

TABLE III. IMAGING SYSTEMS

Dia. (cm)	Type	Typical working range of photon energy (MeV)	Energy discrimination	Intrinsic spatial resolution	Resolving time	Comment	Availability
12.5	Rectilinear scintillation scanner	0.07 – 0.5	Good (e.g. 20 keV at 140 keV)	Limited by collimator only	~2 μs	Resolution limited by collimator and overall speed of mechanical scan system	Commercial
30	Anger camera	0.08 – 0.5	Good (e.g. 20 keV at 140 keV)	~7 mm (150 keV)	~3 μs	Resolution required limits detector thickness and the possible range of photon energies	Commercial
30	Digital autofluoroscope (recent version)	0.08 – 0.5		3 min (static image)	~2 μs	Resolution limited by collimator; high count-rates possible	Commercial
20	Multi-detector positron camera	0.5	Good	~10 mm	<1 μs	Particularly suitable for dynamic studies with short-lived β^+ emitters	Chiefly research Some commercial models
	Image intensifier	e.g. 0.04 – 0.15	Triggering possible	~3 mm	~5 μs	Difficulties with overall sensitivity and energy discrimination being overcome slowly	
	Spark chamber	0.03 – 0.08	Not applicable	~2 – 3 mm	~3 ms	Limited to ^{125}I, ^{197}Hg and no control on detection of scattered photons	
~7	Semi-conductor scanner	~0.05 – 0.15	Very good (e.g. 4 keV at 140 keV)	Limited by collimator only	~1 μs	Good control on scattered photons but limited by detector size, cost and life	Research
5	Semi-conductor camera			~3 mm			

Note: This table is intended as a short summary of the current situation. Details of sensitivity are not readily available in convenient form for comparison in a simple table.

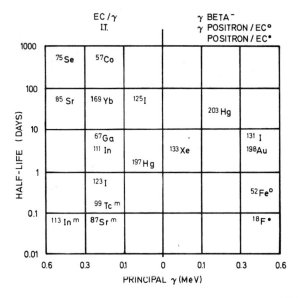

FIG.1. *Chief gamma-ray or X-ray emitting radionuclides used in imaging (1974).*

TABLE IV. RADIONUCLIDE DATA

Energy emitted and half-lives

Nuclide	Energy emitted,[a] Δ (g·rad/μCi·h)			Half-life
	Non-pen	Pen	Non-pen Pen	
^{18}F	0.5157	2.1116	0.25	1.67 h
^{52}Fe	0.4364	1.5640	0.28	8.2 h
^{57}Co	0.0507	0.2564	0.19	270 d
^{67}Ga	0.0776	0.3851	0.20	3.25 d
^{75}Se	0.0420	0.8168	0.051	120 d
^{87}Sr^m	0.1449	0.6757	0.21	2.83 h
^{99}Tc^m	0.0362	0.2675	0.14	6 h
^{111}In	0.098	0.852	0.12	2.8 d
^{113}In^m	0.2774	0.5601	0.49	1.67 h
^{123}I	0.0602	0.3663	0.16	13 h
^{131}I	0.4135	0.8041	0.51	8.0 d
^{169}Yb	0.1520	0.6848	0.22	32 d
^{198}Au	0.7028	0.8631	0.81	2.7 d
^{197}Hg	0.1646	0.1432	1.2	2.7 d

Non-pen: $e^-\, e^+$ and X, γ energy < 0.015 MeV
Pen   : X, γ energy ≥ 0.015 MeV

[a] Basic data derived chiefly from reports of Dillman [11] and Dillman and Von der Lage [12]. See also reference [13].

TABLE V. ABSORBED DOSE AND 'EFFICIENCY' IN CERTAIN SCANNING PROCEDURES[a]

Nuclide	Chemical form	Activity administered i.v. (mCi)	Organ examined	Absorbed dose (mrad)		Length of examination (h)	'Efficiency'[b]	
				Organ examined	Whole body		Assumed effective $T_{\frac{1}{2}}$ for whole body (h)	% of total dose delivered during examination
$^{99}Tc^m$	Pertechnetate	5	Brain	30	65	0.5	5.3	6
^{197}Hg	Chlormerodrin	0.5	Brain (kidney)	– 9300	250	0.5	24 (90%) 65 (10%)	1.5
^{18}F	Complex	1.5	Skeleton	300	110	1	1.8	30
$^{87}Sr^m$	Chloride	2	Skeleton	200	20	1	2.8	20
^{85}Sr	Chloride	0.1	Skeleton	4400	2300	1	2.4 d (60%) 65 d (40%)	0.12
$^{99}Tc^m$	MAA	1.5	Lung	290	2.6	0.5	6	6
^{131}I	MAA	0.3	Lung	1900	24	0.5	5.8 (90%) 72 (10%)	3
$^{99}Tc^m$	Colloid	2	Liver	680	32	0.3	6	3
$^{113}In^m$	Colloid	2	Liver	900	19	0.3	1.8	10
^{67}Ga	Citrate	2	e.g. Spleen	3800	500	~0.5	18 (17%) 78 (83%)	~0.5

[a] These data are based on those of Trott et al. [15] and their references except for ^{67}Ga which is taken from Saunders et al. [16].
[b] These are rough estimates of 'efficiency' in terms of dose delivered to the whole body during examination compared with the total dose resulting from the administration.

Note: This analysis is only intended as a general guide and a detailed study would take full account of delays before the examination starts and the distribution of dose to different organs. In addition, no consideration is given here of the efficiency of the detecting system in recording the quanta emitted, a matter of great importance in comparing, for example, radionuclide and X-ray computerized axial tomography [17].

TABLE VI. ORGAN, BONE MARROW AND GONAD DOSE IN CERTAIN X-RAY (X) DIAGNOSTIC AND RADIOISOTOPE IMAGING (RI) PROCEDURES
(Based on Trott et al. [15])

Technique		Skin dose or activity	Organ investigated	Dose (mrad)		
				Organ investigated	Bone marrow	Gonad
X	Angiography	1000	Brain	(200)^a	130	0.06
RI	$^{99}Tc^m$-pertechnetate scan	5 mCi		30	(65)^b	60 m, 90 f
X	Dorsal spine	2000	Skeleton	(150)^a	200 m 220 f	6 m, 12 f
RI	^{18}F scan	1 mCi		180	(180)^c	73
X	Bronchogram	200	Lung	(50)^a	31	5 m, 17 f
RI	$^{99}Tc^m$-MAA scan	1.5		290	20	10
X	Abdominal radiography	1400	Abdomen	(450)^a	120 m 130 f	100 m 210 f
RI	$^{99}Tc^m$-colloid	2 mCi	Liver	680	38	38 m 46 f

^{a} Estimated dose to centre of organ based on depth-dose data.
^{b} Assumed equal to whole-body dose.
^{c} Assumed equal to dose to skeleton.

physical half-life which will result in the lowest possible radiation dose for a particular type of examination. Thus, if we assume that the examination commences at time t_1 after administration then $\tilde{A} = (A_1/\lambda) \exp \lambda t_1$ where $0.693/\lambda$ is the effective half-life of the administered radionuclide and A_1 is the activity in the patient when the examination commences. If A_1 and t_1 are fixed then the minimum value of $\tilde{A}$ is obtained when $\lambda t_1 = 1$, or in other words the effective half-life should be comparable with the period before the examination commences [8, 14]. It is immediately evident that very short half-life radionuclides can only be used to reduce the dose to the patient for a given quality of image if the time before the examination commences (t_1) can also be reduced. This reasoning will still apply for constant infusion techniques, but of course then a very high proportion of the dose will be given during the test itself.

The problem of physical half-life can also be considered in terms of the 'efficiency' of particular tests, this being defined as the ratio between the dose delivered during the examination to the total dose received before all the radionuclide is cleared from the body. Illustrations of this concept are given in Table V.

For comparison, the dose received by patients in a number of radionuclide procedures and in the corresponding X-ray examinations are given in Table VI.

1.4. Absorption of radiation

Radiation absorption influences the choice of radionuclide in two ways: (a) absorption in tissue, resulting in a reduction of the number of photons reaching the detector and (b) absorption in lead or other high atomic number material used in collimators, the photon energy determining

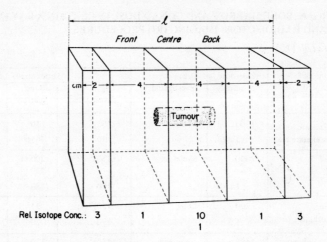

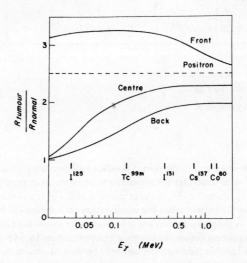

FIG.2. Response over "tumour" relative to normal for gamma rays of various energies and positron annihilation radiation. Tumour viewed by focusing collimator from left side. (Figure 3 in Ref. [5] by permission of the author and publishers.)

the minimum thickness of septa that can be used. The number of photons transmitted without any loss of energy is reduced to 50% by approximately 5 cm of tissue at 0.1 MeV, by 2 cm at 0.035 MeV and 10 cm at 1 MeV. Brownell [5] has investigated theoretically the effect of such tissue absorption for a model including a 4-cm-long cylindrical tumour with a concentration 10 times that of the medium immersed at varying depths in a 16-cm-long phantom, for various photon energies. Results of this analysis are shown in Fig.2. These results clearly illustrate the influence of photon energy and the variation in tumour/background counting ratio that results from various changes in conditions.

$$n = \sqrt{\frac{C_{NT}(1-f)^2 t}{1/KS^2}}$$

where C_T and C_{NT} are radioactive concentrations in target and non-target areas ($\mu Ci \cdot cm^{-2}$).
- t = time interval during which counts are accumulated
- f = C_T/C_{NT}
- $n = \dfrac{\Delta r}{\sigma_{(r)}}$ where r is the count-rate over the background and Δr is the increase in count-rate over the target
- K = sensitivity (counts$\cdot$s^{-1}/$\mu Ci \cdot cm^{-2}$) to radiation emitted from whole body plane source
- S = k/K where k = sensitivity (counts$\cdot$s^{-1}/$\mu Ci \cdot cm^{-2}$) to radiation emitted from the target when the collimator is directly over the target area.

FIG.3. *Tumour detection and counting statistics: summary of analysis by Dewey and Sinclair* [3].

1.5. Tumour imaging theory

The general theory of tumour imaging with radionuclide scanning systems probably originates from the paper of Dewey and Sinclair [3] who developed expressions for the concentration of radionuclide required in a tumour, relative to that in surrounding tissue, needed to permit successful imaging for a particular detecting system. Such studies have been taken up in many institutions, perhaps most extensively in Chicago, Boston and Aberdeen, by Beck, Brownell and Mallard respectively and their colleagues. These studies have given us a clearer understanding of empirical observations of the performance of imaging systems, and have guided improvements in instrument design. Figure 3 summarizes the chief features of Dewey and Sinclair's analysis which lead to the derivation of a figure of merit for an imaging system. Their expression may be given in the simplified form

$$t \simeq \frac{1}{r} \frac{n^2}{(1-f)^2}$$

where t = time interval during which counts are accumulated.
- $n = \Delta r/\sigma$ where Δr is the increase in counting rate over the target and σ the standard deviation in the count-rate over the corresponding non-target area.
- f = ratio of concentration of radionuclide in target to that in non-target area.
- r = count-rate for the detector over the uniform non-target area.

or $\sqrt{N} \simeq \dfrac{n}{f-1}$ where N is the number of counts collected in time t over the non-target area.

If one uses a more rigorous definition of σ, as the standard deviation in the difference of the count-rates over non-target and target areas, then we obtain

$$\sqrt{N} \simeq \frac{n\sqrt{f+1}}{f-1}$$

In this analysis it is assumed that the sensitivity of the collimated detector to a given concentration of activity in the target area equals that for the same concentration in a large plane source, i.e. $S = 1$ (see Fig. 3). In practice $S < 1$ and values of n obtained will be approximately S times those shown, i.e.

$$\sqrt{N} = \frac{n\sqrt{2 + S(f-1)}}{S(f-1)}$$

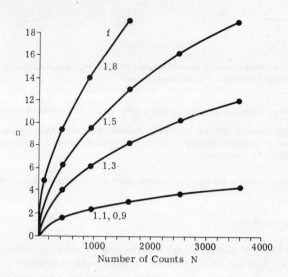

FIG.4. *Counting statistics: significance of observed variations in counting rate (for terminology see Section 1.5). The relation plotted is* $n = \dfrac{f-1}{\sqrt{f+1}} \sqrt{N}$

Mallard and his group have considered in detail the statistical problems of imaging regions of varying contrast (Mallard, Wilks, Corfield and Flook [18]). The influence of changes in f and N can be examined using these expressions and some data are given in Fig.4.

1.6. Effects of scatter

As indicated earlier one limitation on the geometrical resolution that can be achieved arises from Compton scattering in the tissues. This has been analysed by, for example, Eichling, Ter-Pogossian and Rhoten [19] and by Beck et al. [20]. Eichling et al. measured the counts resulting from radiation originating outside the focal region and showed that pulse height discrimination of scattered radiation is "very plausible for ^{131}I and ^{203}Hg, it is more difficult for ^{99}Tc^m, rather inefficient for ^{197}Hg and futile for ^{125}I." They concluded, however, that in practice the variation in effect of scattered radiation on imaging of a model brain tumour was very small over the energy range 77 keV (^{197}Hg) to 279 keV (^{203}Hg), using a scintillation scanner.

2. INTERPRETATION OF CLINICAL OBSERVATIONS

We have examined some of the data obtained on patients by using a gamma camera linked to a multichannel analyser or digital computer in investigations with ^{99}Tc^m-pertechnetate, ^{99}Tc^m-sulphur colloid and ^{67}Ga-citrate, and have tried to relate these to parts of the general analysis of the statistics of image formation. Matters first considered were the areas actually used in reporting scans, the counts recorded in these areas and the activities required to produce the observed counts. We also tried to estimate the concentration of activity in the region examined.

The various observations, including some phantom measurements, cannot be presented in detail. However, Table VII indicates the general magnitude of the data obtained. The concentrations estimated by in-vivo counting are not grossly inconsistent with values expected from biopsy assays.

One particular point of interest, not always appreciated, is the very small activity usually being imaged in locating a tumour, and the correspondingly low concentration of activity in that region of the body. In the ^{67}Ga work, for example, the activity imaged at the tumour site is less than 3% of the activity administered. Thus, apart from the overall inefficiency of the procedures as discussed earlier, there is the additional effect of the general distribution of activity in the body. Nevertheless, if one examines the count-rates over normal and tumour regions, in the light of Fig.4, it will be clear that the differences observed are highly significant, values of n generally exceeding 7. It is equally evident, when one examines the details of the scans, that one is seldom viewing an abnormality against a uniform background; if the difference in counts is close to the statistically significant limit on a simple analysis, only an equivocal report will normally be justified.

One may relate these observations and Fig.4 in another way. Evidently, if activity is already widely distributed in the tissues, and the tumour is small, even a large increase in concentration ratio will only have a modest effect on the count-rate. Nevertheless, doubling the concentration for a given "background" count-rate may be more effective in rendering a tumour detectable than doubling the total activity administered, or the sensitivity of the counting system.

The problem of detecting tumour in a varying background remains severe. The method used by Popham et al. [24] in establishing a recognizable pattern of normals for comparison may only be applicable to the skull and have no place in soft tissue examination. On the other hand, instead of trying to establish normal topographical patterns, it may prove more fruitful to examine temporal ones, using modern quantitative imaging techniques. These can obviously never be so attractive to the clinician and patient as imaging techniques involving only a short single attendance by the patient, but in view of the effort that must go into improving the instruments and pharmaceuticals they should not be overlooked.

3. BIOLOGICAL ASPECTS

The development of the ideal compound can be approached either by studying the physiology and pathology of tumours or lesions or, alternatively, by trying to discover the concentrating mechanisms of compounds in current use.

3.1. Morphology

The appearance of tumours ranges from being very like those tissues from which they are derived to anaplastic types where they are composed of sheets of cells each with similar appearance. Perhaps the only histological feature which might be used to "label" tumours would be to use the difference of cellularity, i.e. concentrations of nuclear material (DNA) between the tumour and normal tissue. The usual DNA labels do not lend themselves to in-vivo imaging methods and in any case the difference in DNA concentration would probably have to be of the order of $5-10:1$ to be useful for "imaging". It is doubtful whether this situation ever exists [25]. Generally, the morphological changes are so slight that considerable experience is required to recognize them. Thus, there is little hope that the variation in DNA concentration between tumours and normal tissue per se could be of value.

One is left then with the possibility of using some feature of tumour metabolism, physiology or cell kinetics. Most of these aspects are still relatively little understood.

TABLE VII. DATA OBTAINED IN TYPICAL GAMMA-CAMERA IMAGING PROCEDURES, WITH ANALYSIS

Material and site of investigation	Activity admin. i.v. (mCi)	Time before imaging	Period of imaging	Area of abnormality, A (cm²)	Total counts in area 'Normal' N	Total counts in area 'Abnormal'	f	Estimated[a] activity in area viewed, N (µCi)	Estimated (N) mean tissue concn (µCi/g)	t^b	d^b (cm)	R
^{67}Ga-citrate (Ca bronchus)	2	48 h	300 s	40	7000	9000	1.3	25^d	0.04	16	5	2
				120	17000	26000	1.5	60^d	0.03	16	8	2
^{99}Tc^m-colloid (liver)	4	15 min	45 s	10	16500	6500	0.4	56^e	0.7	8	2.5	−0.8^c
				3	9170	6610	0.7	32^e	1.3	8	1.5	−0.5^c
^{99}Tc^m-pertechnetate (brain)	8	25 min	300 s	4	2000	2500	1.3	3.2^f	0.057	14	1.5	4
				4	3500	5400	1.6	5.6^f	0.10	14	1.5	7
		40 min	160 min	2	1080	1400	1.3	1.0^e	0.036	14	–	5

[a] Corrected for gamma-ray absorption, assumed 50%, relates to "normal" area N.
[b] t signifies a reasonable value for body or organ thickness. d is calculated on the assumption that the lesion is spherical, i.e. $d = \frac{4}{3}\sqrt{\frac{A}{\pi}}$.
    The effective fluence with relative concentration R in volume dA was calculated as indicated in the note to Fig.5.
[c] To give R = 0 the tumour thickness (d) should be 5 and 2.4 cm respectively (see Fig.5).
    Data based on measurements by W. Guratana [21], L. Jansson (private communication) and S. Paterson (private communication).
    The calibration factors (counts/µCi·s) were 2.1 (d), 13 (e) and 4.2 (f) derived from measurements on a large disc source.

Examples of reports in the literature of assays of tissue specimens are:

	Organ	% of administered activity per gram of wet tissue	Injected amount	Calculated concentration for given injection (μCi/g)	Ref.
^{67}Ga	Liver	0.001 – 0.015 (range of values for specimens from 11 patients)	2 mCi	0.02 – 0.3	[16]
^{99}Tc^m	Brain (mouse)	0.4×10^{-4} at 3 h (stated as 0.03%/1% body wt)	8 mCi	0.003	[22]
^{99}Tc^m	Brain (dura-man)	0.5×10^{-3}		0.04	[23]
^{99}Tc^m-colloid	If we assume that 90% of the activity goes to the liver, then the mean concentration expected would be 2 μCi/g. The site examined was in a region giving a count of density.				

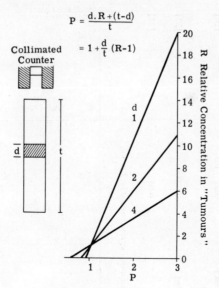

FIG.5. Model to illustrate how change in local concentration of activity affects the total photon output from a cylindrical source for t = 10 cm and d = 1 − 4 cm. Under suitable conditions, namely negligible absorption and compensation for effects of distance, the relative number of photons emitted and so detected by a counter as shown will equal P.

3.2. Cell kinetics

Tumours increase in size at a wide variety of rates [26]. It is not necessarily those that grow most slowly which are benign nor are those which grow most rapidly necessarily malignant. Cellular growth in normal tissue may be very slow or practically absent, for example in the central nervous system or a whole population of cells, for example the gut epithelium, may be renewed every few days [27]. Thus, rapid production of cells with all the associated physiology and metabolism is not a specific feature of tumours. Normal tissues do not increase in size since the cell loss balances the cell production. The growth of tumours is thought to be the result of an alteration in cell production coupled with a change in cell loss. A tumour which doubles its size rapidly might do so not because there is a rapid proliferation of tumour cells but because it is not losing cells [28]. Even if the increase in size were due to tumour cell production alone the speed of growth is small in relation to the turnover of normal gut cells, for example. Doubling times in human tumours have ranged from 10 − 500 days, the medium being about two months [29]. Although the total number of cells may be turned over up to ten times for each tumour doubling volume, still one is not left with the impression that even if a radioactive label which was involved in cell production, such as in a protein or DNA, could be produced it would not be able to show up a tumour against a background of actively dividing normal cells such as is seen in the gut.

Similar arguments could be advanced against the possible use of other radioactively labelled nutrients such as amino acids. That is to say the tumour metabolism is not sufficiently different from normal active tissues to be of value in designing a radiopharmaceutical. A notable exception of course is the metabolism of iodine in functioning thyroid carcinoma. Even this case is rather exceptional in that iodine is stored and converted to hormone rather than used in the production of cells in the tumour. Actual practice however shows that tumours do in fact concentrate amino acids sufficiently to visualize them occasionally.

3.3. Tissue type

In a particular tumour the wide variation in the proportion of tumour cells to other cells has been mentioned already. Such other cells include supporting stroma, vascular endothelial cells and macrophages. When considering tumour localizing agents one is usually subconsciously thinking that such an agent would be localized in the tumour cell component of the mass. However, macrophages have been shown to concentrate the so-called tumour-localizing agent (e.g. gallium) [30], and it has not yet been disproved that this is the main reason for the successful gallium scans. It is known that the macrophage content of tumours varies [31] and it could be that this alone would explain the variation in gallium concentration seen in clinical practice.

It could be that even apart from macrophages the most metabolically active part of the tumour is in fact the non-tumour cells. Thus, a successful localizing agent in these types of tumour would be localized in "normal" cells of the tumour mass.

3.4. Vascularity

It has been shown that the vascular space in tumours may be of the order of three times that of normal supporting tissue [32]. This ratio is of the same order as measured uptake ratios of tumour-localizing agents in human and animal tumours. It could be argued that a simple vascular label would be sufficient to demonstrate such tumours. However, although human tumours have been demonstrated by even ^{51}Cr-labelled red cells [32] and during the initial perfusion of such agents as pertechnetate, not even all so-called vascular tumours show consistent concentration.

The perfusion of tumours is bound to play a part in the concentrating ability of such masses. In general the poorly perfused necrotic areas show less concentration of the known radiopharmaceuticals [33]. It is difficult to know whether the richly perfused areas with apparent higher growth rates which show good radiopharmaceutical localization do so because of the growth or the good blood supply. Thus, one is left with a situation where one is trying to design a compound to localize tumours which may turn over cells more slowly than normal tissue, may have a relatively poor blood supply, and may be composed of perhaps up to 50% of non-tumour cells.

The alternative approach is to study the more successful tumour-localizing radiopharmaceuticals and discover the reason for their success in locating tumour and producing positive scans more frequently in malignant than benign lesions. Since this is being done in detail elsewhere in these Proceedings the following comments will be limited to generalities. So far most research has concentrated on gallium, mercury and indium compounds. It is not really clear why they behave in the way that they do, and often the experimental results are conflicting. For example, gallium will show up most malignant and many benign and inflammatory lesions [34]. It has been detected in tumour cells [35] and macrophages [36]. In the cell it has been reported to be at times both in the cytoplasm [30] and nucleus [37]. It has been associated with the lysosome-like structures [36], tumour membranes and mitochondria [38]. It appears to enter the cell just after DNA synthesis or concentration [39]. It is not obvious in what form it enters the cell but the fact that it enters at G2 similarly to other ions suggests it might be as an ion, but since it is protein-bound in the blood it might also be pinocytosed. However, similar levels of uptake and comparable clinical results are found with most of the other well-known tumour-localizing agents except those directly involved in metabolism (e.g. melanin precursors and iodine). This could suggest a basically simple mechanism such as pinocytosis of passively labelled protein [40] (perhaps with the exception of ^{111}In-labelled or ^{57}Co-labelled bleomycin and labelled antibiotics). However, even in the latter cases there is no evidence that the compounds are concentrated in the cell as the intact labelled compound. Since antitumour agents have not been shown to concentrate in tumours there is no reason to suppose that labelled antitumour agents should be more successful except in the special case of animal brain tumours [41]. Thus, even with the known tumour-localizing agents it is difficult to find the mechanism which might be used to prepare the ideal tumour-localizing agent let alone an agent which reflects tumour actual growth.

In the end we are left with the empirical method of finding the ideal compound, its choice relying more on good physical characteristics and less on concrete biological reasoning. Until we know more about the chemistry and physiology specific to malignant tumours the biological reason for choosing a compound will still have to rely more on intuition than logical deduction.

ACKNOWLEDGEMENTS

Numerical data on liver scans and some brain scans were obtained by Mr. Lars Jansson using a PDP8 system, developed by Dr. R.E. Bentley, connected to an H.P. Nuclear Chicago gamma camera. Other data were obtained by Dr. Wanida Guratana and Dr. S. Paterson using an Intertechnique Tridact data analysis system connected to a Nuclear-Chicago Pho/Gamma III.

REFERENCES

[1] PATERSON, A.H.G., McCREADY, V.R., Tumour imaging radiopharmaceuticals: a review, Br. J. Radiol. **48** (1975) 520.
[2] WILLIS, R.A., Pathology of Tumours, Butterworths, London (1967).
[3] DEWEY, W.C., SINCLAIR, W.K., Criteria for evaluating collimators used *in vivo* distribution studies with radioisotopes, Int. J. Appl. Radiat. Isot. **10** (1961) 1.
[4] BECK, R.N., "A theory of radioisotope scanning systems", Medical Radioisotope Scanning (Proc. Symp. Athens, 1964) **1**, IAEA, Vienna (1964) 35.
[5] BROWNELL, G.L., "Theory of radioisotope scanning", Medical Radioisotope Scanning (Proc. Symp. Athens, 1964) **1**, IAEA, Vienna (1964) 3.
[6] MATTHEWS, C.M.E., Comparison of isotopes for scanning, J. Nucl. Med. **6** (1965) 155.
[7] HARPER, P.V., BECK, R., CHARLESTON, D., LATHROP, K.A., Optimization of a scanning method using ^{99m}Tc, Nucleonics **22** (1964) 50.
[8] MALLARD, J.R., PARKER, R.P., TROTT, N.G., "Use of shortlived radionuclides in high spatial resolution studies in nuclear medicine", Int. Conf. Peaceful Uses Atom. Energy (Proc. Conf. Geneva, 1971) **13**, UN, New York (1972) 85.
[9] HARPER, P.V., "Physical and nuclear parameters", Ch. 1, Fundamental Problems in Scanning (GOTTSCHALK, A., BECK, R.N., Eds), C.C. Thomas, Springfield, Ill. (1968) 5.
[10] MAGEE, B., Popper, Fontana/Collins, London (1973) 32.
[11] DILLMAN, L.T., Radionuclide Decay Schemes and Nuclear Parameters for Use in Radiation Dose Estimation, Pamphlets 4 and 6, Society of Nuclear Medicine, New York (1969, 1970).
[12] DILLMAN, L.T., Von der LAGE, F.C., Radionuclide Decay Schemes and Nuclear Parameters for Use in Radiation Dose Estimation, Pamphlet 10, Society of Nuclear Medicine, New York (1974).
[13] TROTT, N.G., O'CONNELL, M.E.A., ROSS, H.A., SMITH, P.H.D., TAYLOR, D.M., "Some studies of the dosimetry and safety of radiopharmaceuticals", Radioaktive Isotope in Klinik und Forschung, Vol. 11 (FELLINGER, K., HÖFER, R., Eds), Urban and Schwarzenberg, Munich (1974) 1.
[14] WAGNER, H.N., EMMONS, H., Design of new radiopharmaceuticals *in* Symposium on Computers and Scanning (HIDALGO, J.U., Ed.), Soc. Nucl. Med., New York (1967) 57.
[15] TROTT, N.G., STACEY, A.J., ELLIS, R.E., DERMENTZOGLOU, F., "The dosimetry of selected procedures using X-rays and radioactive substances", Medical Radionuclides: Radiation Dose and Effects (CLOUTIER, R.J., EDWARDS, C.L., SNYDER, W.S., Eds), USAEC, Symposium Series No. 20, Oak Ridge (1970) 157.

[16] SAUNDERS, M.G., TAYLOR, D.M., TROTT, N.G., The dosimetry of ^{67}Ga citrate in man, Br. J. Radiol. **46** (1973) 456.
[17] KEYES, W.I., Radioisotope section scan versus E.M.I. scan, Br. J. Radiol. **45** (1975) 1036.
[18] MALLARD, J.R., WILKS, R.J., CORFIELD, J.R., FLOOK, Valerie, "Visualization in scanning", Medical Radioisotope Scintigraphy (Proc. Symp. Salzburg, 1968) **1**, IAEA, Vienna (1969) 305.
[19] EICHLING, J.O., TER-POGOSSIAN, M.M., RHOTEN, A.L., "Analysis of the scattered radiation encountered in lower energy diagnostic scanning", Ch. 19, Fundamental Problems of Scanning (GOTTSCHALK, A., BECK, R.N., Eds), C.C. Thomas, Springfield (1968) 238.
[20] BECK, R.N., ZIMMER, L.T., CHARLESTON, D.B., HOFFER, P.B., LEMBARES, N., "The theoretical advantages of eliminating scatter in imaging systems", Ch. 7, Semiconductor Detectors in the Future of Nuclear Medicine (HOFFER, R.N., BECK, R.N., GOTTSCHALK, A., Eds), Society of Nuclear Medicine, New York (1971) 92.
[21] GURATANA, W., Investigation with a gamma camera of the uptake of ^{99m}TcO$_4$ in normal brain and brain tumours, M. Sc. Report, University of London (1971).
[22] McAFEE, J.G., FUEGER, C.F., STERN, H.S., WAGNER, N.J., Jr., MIGITA, T., ^{99m}Tc pertechnetate for brain scanning, J. Nucl. Med. **5** (1964) 811.
[23] MISHKIN, F.S., REESEN, T.C., Tissue and tumour concentration of ^{99m}Tc as pertechnetate, Am. J. Roentgenol. **104** (1968) 145.
[24] POPHAM, M.G., BULL, J.W.D., EMERY, E.W., Interpretation of brain scans by computer analysis, Br. J. Radiol. **43** (1970) 835.
[25] HAMMERSLEY, P.A.G., TAYLOR, D.M., ^{67}Ga citrate incorporation and DNA synthesis in tumors, Ch. 47, Radiopharmaceuticals (SUBRAMANIAN, G., RHODES, B.A., COOPER, J.F., SODD, V.J., Eds), Soc. Nucl. Med., New York (1975) 447.
[26] CHARBIT, A., MALAISE, E.P., TUBIANA, M., Relation between the pathological nature and the growth rate of human tumours, Eur. J. Cancer **7** (1971) 307.
[27] STEEL, G.G., "Cytokinetics of neoplasia", Ch. II-2, Pathogenesis of Cancer, Lea and Febiger, Philadelphia (1973) 125.
[28] STEEL, G.G., Cell loss from experimental tumours, Cell Tissue Kinet. **1** (1968) 193.
[29] STEEL, G.G., LAMERTON, L.F., The growth rate of human tumours, Br. J. Cancer **20** (1966) 74.
[30] HAYES, R.L., NELSON, B., SWARTZENDRUBER, D.C., CARLTON, J.E., BYRD, B.L., Gallium-67 localisation in rat and mouse, Science **167** (1970) 289.
[31] EVANS, R., Preparation of pure cultures of tumour macrophages, J. Natl. Cancer Inst. **50** (1973) 271.
[32] McCREADY, V.R., Unpublished observations.
[33] NASH, A.G., DANCE, D.R., McCREADY, V.R., GRIFFITHS, J.D., Uptake of gallium-67 in colonic and rectal tumours, Br. Med. J. **3** (1972) 508.
[34] ITO, Y., OKUYAMA, S., AWANO, T., TAKAHASHI, K., SATO, T., KANNO, I., Diagnostic evaluation of Ga scanning of lung cancer and other diseases, Radiology **101** (1971) 355.
[35] ITO, Y., OKUYAMA, S., SATO, K., TAKAHASHI, K., SATO, T., KANNO, I., Gallium-67 tumour scanning and its mechanism studied in rabbits, Radiology **100** (1971) 357.
[36] SWARTZENDRUBER, D.C., NELSON, B., HAYES, R.L., Gallium-67 localisation in lysosomal-like granules of leukaemic and non-leukaemic murine tissues, J. Natl. Cancer Inst. **46** (1971) 941.
[37] BICHEL, P., HANSEN, H.H., The incorporation of gallium-67 in normal and malignant cells and its dependance on growth rate, Br. J. Radiol. **45** (1972) 182.
[38] HIGASI, T., NAKAYAMA, Y., AKIBA, C., HISADA, T., FIELDS, M.H., The mechanism of uptake of gallium-67 in tumour cells, Radioisotopes **22** (1973) 291.
[39] MORRANT, M., McCREADY, V.R., Unpublished observations.
[40] BUSCH, H., SIMBONIS, S., ANDERSON, D.C., GREENE, H.S.N., Studies on the metabolism of plasma proteins in tumour bearing rats. II. Labelling of intracellular particulates of tissues by radioactive albumin and globulin, Yale J. Biol. Med. **29** (1956) 105.
[41] HAYAKAWA, T., USHIO, Y., MOGAMI, H., HORBATA, K., The uptake distribution and anti-tumour activity of Bleomycin in gliomas in the mouse, Eur. J. Cancer **10** (1974) 137.

DISCUSSION

R.M. KNISELEY: Can you tell us about the present sensitivity and resolution of ultrasound techniques?

V.R. McCREADY: Ultrasound cannot find tumours in all parts of the body, as ultrasound won't go through air-containing structures. The resolution of ultrasound is down to about 3 mm, at best, inside the liver. The fact is that there are multiple areas in the body, e.g. lungs, brain and other organs where radionuclides should be better than ultrasound, both for physical and functional reasons.

W.H. BEIERWALTES: I have asked various hospitals throughout the United States, where they have carried out studies for an extended period with ultrasound and EMI scanners (or CAT scanners): What happens to brain tumour scintigraphy in a place with a good EMI scanner? There has been a temporary decrease in the utilization of brain scans, for instance at the Mayo Clinic, but later it plateaued with no further rise or fall to about the level of brain scintigraphy before the EMI scanner was introduced. The EMI scanner and ultrasound measure primarily structure, whereas radionuclides, at least at their best, describe function in addition to structure. The resolution of an EMI scanner is degraded if the patient moves, and good EMI studies depend on the patient's co-operation (or the administration of deep anaesthesia).

Not all tumours persist after cessation of the stimuli which have evoked the tumours. An endocrine-dependent breast carcinoma may regress after oophorectomy. A radioactively "hot" adenoma in the thyroid gland responsive in fuction to T_3 suppression may disappear completely, at least as long as T_3 or thyroid hormone substitution is continued.

It is very important to detect small tumours; in fact we should try to detect disease before it has produced symptoms. Schilling's test is a good example of demonstrating a biochemical defect before the patient has shown any signs or symptoms of disease. It is our experience that the majority of phaeochromocytomas and hyperfunctioning adrenocortical adenomas causing Cushing's syndrome have a diameter of more than 3 cm at the time of diagnosis and treatment. We believe that there is a much larger population of smaller tumours causing symptoms, but they are not yet being diagnosed by other methods. If we can achieve a high target-to-non-target ratio it may be possible to detect disease before it has produced symptoms. The most outstanding example is of course the radioimmunoassay for serum thyrocalcitonin which can detect medullary thyroid carcinoma before any goitre symptoms have occurred. If the serum thyrocalcitonin rises after pentagastrin stimulation, it is possible that you don't find a medullary thyroid carcinoma, but you will always find the precursor to the carcinoma, namely C-cell hyperplasia. One of the greatest advances in nuclear medicine will be to detect tumours at the precursor stage before the neoplasm is actually present.

As to tumour detection based on metabolic abnormalities, it is somehow depressing that the biochemical characteristics of normal tissues disappear when the tissues become cancerous. Tumours seem to sacrifice function for growth. But, in analogy with bacteria which may develop specific traits under unfavourable environments, tumours might also develop specific enzyme activity during growth which we might detect with radiolabelled enzyme inhibitors. Recently, we have been successful in achieving diagnostic concentrations of radiolabelled enzyme inhibitors in the adrenal cortex.

I hope that by the time I retire we will be able to image tumours 2 mm in diameter at any depth in the human body. For the moment we can lateralize altosteronomas $5 - 8$ mm in diameter, but as a general rule we are only discretely imaging these tumours when they are larger than 2 cm in diameter.

I should like to stress our experience in bone scanning with technetium diphosphonate or pyrophosphate where the gamma camera has been superior to the rectilinear scanner in sharply defining and accentuating the contrast in areas of metastates. We start with a total-body scan, and then, in case of a suspicious area, we continue with a focal gamma-camera study of the lesion in question.

In our experience the use of a computer has only been of benefit in two situations: first, in making it possible to alter the display of adrenal imaging by means of ^{131}I-19-iodocholesterol to enhance its diagnostic use, and second for calculating the per cent uptake of the iodocholesterol in the adrenals.

As to the comments on loss of photons due to low energy and intercalated tissue, I want to mention that it may sometimes be useful to apply low-energy radionuclides, under certain definite conditions. We have been able to use ^{131}Cs for myocardial scintigraphy in patients with infarcts, and we have been able to show metastases of malignant melanoma with an ^{125}I-chloroquine analogue in the liver in the human, certainly close to the surface.

As to the speed by which we get the radiolabelled compound into the tumours, we have shown an uptake in the adrenals of ^{14}C-oestradiol 10 min after administration and a maximum of ^{14}C-diphenylhydantoin in the islet cell of pancreas 10 min after administration. These observations that we can have a high target-to-non-target ratio 10 min after administration will make it possible to use ^{11}C compounds if we can solve the problems of rapid radiochemistry. In studies with ^{14}C-dopamine in the adrenal medulla we obtained a good target-to-non-target ratio within 2 to 4 hours, and the studies showed that 96% of the radioactivity was excreted in the urine within 6 hours after administration and 50% of the radioactivity had disappeared from the blood within 20 min.

For years we have treated metastatic thyroid carcinoma successfully with radioiodine, and I should like to stress that we never achieve a concentration in the metastases higher than 0.28% of the dose per gram. In imaging the adrenals by means of iodocholesterol we have obtained 0.6% of the dose per gram in the adrenals and in the tumours, and when we started with our first irreversible enzyme inhibitor labelled with ^{14}C, we achieved 8% of the dose per gram uptake in the adrenals of rats. Finally, we have obtained 20% uptake of dose per gram in mouse melanomas when using a new iodochloroquine analogue. It is not always easy to predict the uptake of the labelled compounds in tumours. It has been shown that the phosphorous content of the nuclei of leukaemic cells was not higher than in nuclei of lymph nodes. The concentration of ^{32}P, however, was ten times higher in the nuclei of leukaemic cells. This may have to do with the reduplication of cells. The highest rate of replication is in certain cells in the hair follicles. Some cancer chemotherapy agents also accumulate in these cells. ^{131}I human serum albumin has been used by us as a reference for vascularity of tumours, but we have also found that it was concentrated four times more in malignant melanomas than in blood. As to antitumour agents, I think you are right that they are normally not concentrated in tumours. I have, however, seen a recent report where ^{14}C-methotrexate was concentrated in a glioma with a maximum after 60 min, but unfortunately it was only demonstrated by autoradiography.

H.J. GLENN: First, I should like to point out that when dealing with an ^{11}C compound we are not only hampered by the fast chemistry, but we must also be able to do a fast quality control of the products. If you don't have a strict control of the radiopharmaceutical, your results may be meaningless.

Another point of interest is that you probably will find a limit as to the chemical mass which the tumour or tumour components can accumulate. We have used a melanoma cell-line for many years as our standard and we have studied the gallium uptake in the tumour cells. We found that the added carrier had a pronounced effect, but we also found that we had non-radioactive gallium in the compound which we used because the supplier had used his target for the first time. This means that inactive gallium was present in the zinc used for the cyclotron preparation. In this way our technique was sensitive enough to pick up carrier gallium in what was supposed to be carrier-free material. When you want to synthesize an iodocholesterol, you have to work with enough material for its chemical production and this may limit the use of the final compound. In this area carrier-free materials are an advantage.

As to a reference for the vascularity of tumours, we have used ^{51}Cr-labelled red cells to estimate tumour blood concentration rather than iodinated albumin because iodinated albumin does penetrate some tumours to a certain extent.

T. MUNKNER: By means of neutron activation analysis we have found that the ^{67}Ga preparations available in Europe have an isotopic abundance close to 100%. Preparations from other suppliers may be different. The concentration of stable gallium was found to be extremely low in normal tissues, and the concentration was not increased in malignant tumours.

W.H. BEIERWALTES: Were these studies carried out on homogenates, on cells, or on subcellular fractions?

T. MUNKNER: The concentrations of stable gallium were determined in tumour tissues (which were verified microscopically) and in tissues just outside the tumour areas. The studies were not carried out on cell fractions.

R.L. HAYES: The presence of stable gallium in ^{67}Ga preparations depends on whether the manufacturer uses isotopically enriched zinc as target material. Such preparations of ^{67}Ga should be quite clean since the target material is reused. There is always the possibility, depending on the way the manufacturer makes the separation, that there might be some stable zinc present in the ^{67}Ga preparation. It is almost impossible to keep out a small amount of iron − I am talking about less than microgram quantities − but it can have an influence on the preparation.

There have been some recent reports in the Russian literature about the measurement of stable gallium concentrations, not specifically in tumour tissues, but in a wide variety of diseases, and it was my impression that it had no particular correlation with anything.

T. MUNKNER: Gallium was measured years ago in Russia in different tissues. I have the impression that gallium is used as some part of the teeth fillings in Russia. If so, this may change the whole story.

R.L. HAYES: There has been considerable work on the use of gallium as an amalgam.

H.J. GLENN: Basically, it is only the first time the target is used that the suppliers will run into the problem with natural gallium impurities that are present in zinc. By the gallium chemical separation from a new target, they will remove all of the non-radioactive gallium impurities in the zinc as well as the radiogallium. By recycling the target material − and I think that most of the commercial suppliers use recycled gallium, − essentially carrier-free material results.

D. COMAR: If nobody else will do it, I will defend the work done with cyclotron-produced radionuclides. Short-lived radioisotopes made by cyclotron are not only used for tumour localization. That means that the cyclotron will not be eliminated from nuclear medicine, even if it doesn't produce useful short-lived tumour-localizing agents. In addition, it is not always possible to draw conclusions from studies carried out with ^{14}C-labelled molecules, where the specific activity is bound to be rather low, to studies carried out with the same compound but labelled with ^{11}C, which can be given in much smaller quantities (10^{-9} to 10^{-11} times the amounts of the ^{14}C compounds).

W.H. BEIERWALTES: My point was that short-lived radionuclides with a half-life of less than one or one and a half hours don't enjoy a very popular use today in routine clinical medicine. I think there will be a great future for cyclotron-produced radionuclides when the chemistry can be accomplished to radiolabel a compound in perhaps 10 min, and when we find radionuclide-labelled compounds that concentrate in high per cent dose per gram and in high target-to-non-target ratios in a very short time. In our work on the comparative uptake of oestrogens and androgens in the dog prostate we used compounds with relatively high specific activities (around 49 mCi per millimole of ^{14}C and about 59 Ci per millimole of the tritiated material).

V.R. McCREADY: My presentation was focused on localizing or, if you want, visualizing agents, and I did not include comments on other nuclear medicine tests, such as Schilling's test, calcitonin measurements or CEA determinations. It is likely that our treatment of tumours will be a combination of radiotherapy, surgery and chemotherapy, and for this reason there will be a need for pictorial demonstration of the tumours. But obviously we will also follow other lines of approach, if feasible, as for instance mentioned by Dr. Beierwaltes in his remarks on enzymes.

W.H. BEIERWALTES: It is important to stress your comments on getting positive localization of radionuclide-labelled compounds in tumours. Besides the physical reasons for this statement, you may argue that the appropriate approach to radionuclide therapy of tumours is not by lack of uptake in the tumour, but via a positive demonstration of an uptake in the cancerous tissues. If the tumours do concentrate the radiolabelled compound, we should always go on and try therapy for two reasons. First, because therapy might work, and second because these studies may prove that there is absolutely no demonstrable effect of the therapeutic doses on any organ of the body in a long period of time. Such findings would suggest that tracer doses are not creating hazards of ionizing irradiation. In recent studies in the dog under varied conditions, the LD_{50} of ^{131}I-19-iodocholesterol was given without any changes in the adrenals, histopathologically or functionally. Under these circumstances it is difficult to criticize the radiation dose to the adrenals in normal individuals. By the same token, therapy doses of ^{131}I-chloroquine analogues have been given to dogs with dermal melanomas, and we succeeded in curing one dog as shown not only by the growth disappearance of the primary tumour and the metastases, but also by microscopic studies after sacrificing the dog. After these studies we went on and did the same thing in humans. One of the possible harmful side-effects of such a treatment would be the irradiation of the retina by the ^{131}I concentration in the choroid where there is so much melanin. By the most sensitive method of detection, that is photometric examination of the response of the rods and the cones, we were able to show that by giving amounts corresponding to LD_{50} of ^{131}I-chloroquine analogues to numerous dogs and following the visual response for a minimum of three months as well as following nine patients with malignant melanomas treated finally with doses that depressed the bone marrow, there was no detectable visual change. Knowing these results, it would be very difficult to be really concerned about giving tracer doses of iodochloroquine for the detection of ocular melanomas.

V.R. McCREADY: Let me make a comment on the problem of camera versus rectilinear scanner. If we develop tumour-localizing agents, of necessity we want a whole-body survey. The camera is fast whereas the rectilinear scanner is relatively slow. However, the camera images tumours best near the surface whereas scanners usually image deeper structures more effectively. A dual-headed scanner can make the scanning technique less depth-dependent. Then there is the problem of organ motion. This is especially a problem in rectilinear scanning. I think we clinicians should be more helpful and try to define a set of conditions as near ideal as possible to enable physicists to design a system for efficiently visualizing tumours of a given minimum size at a given concentration ratio at $3 - 10$ cm below the skin surface. We should attempt to define the maximum acceptable time for a typical whole-body scan. The hybrid scanning camera may be the answer but, at least in bone scanning, rib lesions visible in camera studies are not seen on rectilinear scans.

W.H. BEIERWALTES: In our experience, too, we have seen metastatic carcinoma in bone beautifully with a camera that we could not demonstrate at all with a rectilinear scanner.

V.R. McCREADY: I repeat that in my view the trouble is that the nuclear medicine physicians have not defined the problem for their physics colleagues. We leave our physics colleagues working with phantoms trying to generalize a situation that is not general. The camera will see spinal lesions better than the rectilinear scanner because we are using collimators for the scanner that usually visualize a block of tissue some depth below the surface, whereas with a cemera you have no option, you will only see the first four centimetres of tissue. In addition, the camera image does better because it is more efficient in terms of information density per unit time.

From the table that I showed it may be concluded that we need specialized scanners for energies between 100 keV and 200 keV, scanning the whole body at a depth of $2 - 6$ cm. In addition, you should consider if you could achieve better results by giving more of a short-lived radionuclide instead of spending the money on more sophisticated machinery. The technique of display is crucial. The only reason that scanners were so impressive over the years was that they made highly contrasted black and white pictures. This has gone now, because we are using digital scanners which produce pictures that are grey and white. Finally, I want to stress that we totally lack information on concentrations in organs as well as in tumours. We cannot compare ^{197}Hg with ^{67}Ga at this time because this information is lacking. For example, there is no point in comparing the accuracy of my series with Dr. Hisada's series because my scanner might be better than his.

W.H. BEIERWALTES: Can you amplify your remarks on the semiconductor detectors?

V.R. McCREADY: In our rather short experience the semiconductor camera had a better resolution than the older-type gamma camera. We had to stop our studies with the semiconductor camera because the people involved in producing it could not continue to do so. The principle is still as good as it ever was. The sensitivity of the semiconductor is down by a magnitude compared with the Anger camera. To take the full advantage of the inherent resolution of a semiconductor camera, you have to make your collimator just as good as the inherent resolution of the detector. If you do that, then you finish up with more lead than hole. This may call for an increase in your radioactivity by a power of maybe four, which takes us back again to short-lived isotopes in very high activities. And you are going to have some very clever way of getting them into the patient without excessively irradiating the operators.

W.H. BEIERWALTES: The organic chemists much prefer ^{131}I. In nuclear medicine we need a resolution of 3 − 5 mm at a depth of 3 − 4 in with a gamma energy of 364 keV.

V.R. McCREADY: This raises the problem of the importance of ^{123}I. This would give us an ideal label if we could reduce the time necessary for labelling. It is really important to decide whether ^{123}I is an isotope worth pushing because if we require cyclotrons we have to plan about four years ahead.

D. COMAR: At some time during our discussion I think we should make the comment that we don't have any physicists around the table. Maybe this is on purpose, but I imagine that there are things to be said on detectors, computers and programmes which could be of great help in tumour detection.

T. MUNKNER: We have had the same experience as mentioned by Dr. Beierwaltes that a dip was noticed in brain scintigraphy when the EMI scanner went into action. We discussed how our methods might change in the near future and how in the diagnosis of brain tumours we might have the possibility of another approach by data processing of dynamic studies. The computers are already available in many departments and dynamic studies might improve the diagnosis. We cannot compete with the EMI pictures by using radionuclide brain scintigraphy. Even if a few EMI pictures are disturbed by movements, the perfect EMI pictures are much superior to brain isotope scans, and the majority of the pictures in my hospital are perfect. Personally, I am reluctant to initiate radionuclide screening of patients with minor signs and symptoms. It will change the patient load in a department of nuclear medicine and the percentage of true positive results will be low.

The patients who are scanned immediately with EMI are those with acute catastrophic symptoms. In the future primarily patients with diffuse symptoms or a long-standing story of vague neurological signs (mostly patients from neuromedical departments) will be sent for radionuclide scintigraphy. The percentage of positive scintigrams will decrease and the diagnostic acumen of the nuclear medicine departments will be challenged much more than they are today.

If parathyroid scintigraphy is to help the surgeon, we must be able to demonstrate tumours of 20, 50 or 100 mg size. In the Middlesex series of parathyroid tumours, one was between 1 and 2 g and all the others were above 2 g. Parathyroid tumours of this size are often called "potatoes" by the surgeons, and every surgeon should be able to find such a tumour by exploration of the neck or mediastinum.

In nuclear medicine we try very hard (and until now in vain) to design methods which give results that conform with the term "malignancy". I think we should try to make our own "nuclear terminology", try to tell what we show and not to fill the frames which have been set by other disciplines. We will never be able to show the invasion into vessels of malignant cells, and so on. I am in favour of a new terminology telling that the result is "gallium-positive" or "gallium-negative", which is exactly what we have demonstrated by our methods.

V.R. McCREADY: When you speak about dynamic studies, do you mean studies where the patient doesn't move from the start to the finish of the examination, or do you mean dynamic studies as they have been carried out by Dr. Planiol, where you study the distribution of a compound over a series of days?

T. MUNKNER: The dynamic studies referred to were measurements, for instance of brain blood flow, where changes can be demonstrated in the case of brain tumours. Measurements of brain blood flow, for instance with ^{133}Xe, can be valuable for the detection and even for the differential diagnosis of brain tumours.

D. COMAR: Dynamic studies, in my opinion, can also be studies extending for a couple of hours. I think, for instance, of the uptake of short-lived isotopes which may be higher in one part of an organ than in another part, such as, for instance, the localization of psychotropic drugs in the brain. Dynamic studies in this case will give you a good idea of the metabolism of the molecules.

V.R. McCREADY: I want to take up Dr. Munkner's second point of putting a new type of label on a disease such as "gallium-positive" and "gallium-negative". I would submit that there are two types of gallium work: there is academic work which is done only for physiological interest, and there is clinical work which must fit in with the needs of the clinician. The clinician is basically interested in finding disease, determining its spread and whether or not it is responding to treatment.

T. MUNKNER: In many fields it is wise to say what you see and not what you think about the object, if you are not 100% certain. I know what the clinicians want. They want to push us into telling whether a hot or cold spot is malignant or not, but that is exactly what we are unable to tell them for the moment.

V.R. McCREADY: Yes, it could be a big step forward if we could relate gallium uptake to a histological diagnosis of malignancy. However, work carried out by Dr. Patterson at Sutton has shown that patients with advanced disease show very high uptake as well as multiple areas of concentration. So obviously the gallium concentration is related, at least grossly, to malignancy in the sense of the disease progressing to death.

W.H. BEIERWALTES: We have asked our clinicians if they want us to tell if there was an increased uptake in an area or not, without any interpretation, or whether they want us to suggest some probabilities and diagnoses. The general response was that they want probability of diagnoses. So now we stage our diagnosis, for instance by saying it is a "code 5", which means definitely abnormal uptake, or for instance "code 1", normal. I think we can frequently tell the clinician a tumour-specific diagnosis. As to specificity of the cancer diagnosis, I can give some examples. If we find a woman who has a clear-cut increased uptake of technetium diphosphonate over the area of the breast and find a series of sharply localized radioactivity concentrations in the region of the pedicles of the vertebral bodies, we have never been wrong in the fact that this was breast carcinoma, and metastatic to bone. Similarly, in a case where we found absolutely no radioactivity in a sharply circumscribed circle in the region of the breast and numerous metastases to the spine, we have never been wrong in diagnosing that this woman is wearing a breast prothesis because she has had a radical mastectomy. Further, if we find uptake in the region of the upper part of the right lobe of the lung, particularly if it involves the rib of that area, and then we find numerous areas of radioactivity in the skeleton, we have never been wrong so far in saying that this was a bronchogenic carcinoma.

R.L. HAYES: Dr. McCready, you very reasonably raised the question whether short half-life radioisotopes could be useful ultimately, and I certainly don't argue about the fact that the short half-life almost makes it essential that you have certain selected sites for this particular type of diagnosis. Obviously, the expenses entailed are tremendous, so you want some sort of reasonable assurance that what you are buying will eventually pay for itself. We are for the moment interested in ^{11}C, and we have a couple of compounds which I want to mention later on when the subject comes up.

I have been told that they are trying in St. Louis to use the EMI technique with positron collimation. From the theoretical standpoint I understand this seems to have considerable potential. We might be on the verge of really utilizing the theoretical potential of the positron and detection of fairly low-level tumour-to-non-tumour ratios might become quite feasible, particularly as one gets rid of the scatter. We might not be confined specifically to positive uptake, but here perhaps is the case where cold lesions might come through very well.

V.R. McCREADY: Certainly the tomographic pictures produced by Ter-Pogossian with positrons and correct absorption calculations are very impressive. I am not sure from a morphological viewpoint that they are better than EMI scans.

R.L. HAYES: What about the pancreas for example. Suppose we are talking about an active agent going to normal pancreas with a void. In such a case would this be productive?

V.R. McCREADY: I do not think that positron scanning would be a great advantage but I really do not have any experience. I have found that this cross-over of ideas in a reasonably limited time pays more dividends than any other way of working. We have found that we do better in ultrasound because we are using nuclear medicine techniques in our ultrasound display. So I think it is quite right to raise this sort of combined way of attacking things because quite fortuitously they often pay off much better than one would have otherwise thought.

E.H. BELCHER: At this point I should like to make a couple of remarks on the physical aspects of imaging techniques, particularly as regards hot and cold lesions. The IAEA has at the present time a co-ordinated research programme which is concerned with the intercomparison of computer-assisted data-processing techniques for imaging, i.e. for image enhancement by computer processing of data. Up to now, this programme has been based on the distribution of simulated scan data which were distributed to participants in the programme, processed by them, and then the results returned to the Agency for evaluation. It has been rather surprising to find that, in fact, cold lesions are slightly easier to detect than hot ones of the same significance. I have to qualify this statement to some extent; this finding applies to situations where one has lesions superimposed in a rather hot background, such as one might have in a liver scan, for example. Perhaps one has to be a little bit cautious in deciding that a radiopharmaceutical which will have an increased uptake in the tumour is absolutely necessarily the best choice. Perhaps there may be situations where this is not true. Our results raise a lot of questions. First of all the questions as to how one should define the significance of the lesion in relation to the background in the neighbourhood of the lesion. It raises various questions of perception, particularly as to how the observer perceives either a hot or a cold lesion. Purely on statistical grounds there may be a basis for this finding. If the observer perceives a lesion by comparing the counting rate over the target area with a counting rate over neighbouring or background area, then the statistical error in the observed difference between these two numbers of counts will in fact be greater if a tumour has the higher rate to the background than if it is the other way round. But, once more, I want to stress that we observed this in the co-ordinated research programme which is based on studies with a very simple geometric phantom, a situation which is very far from the actual clinical situation.

W.H. BEIERWALTES: This problem is an extremely complex topic and you have to qualify under what circumstance you can best detect the cold spot in a certain background and under what circumstances you can detect the hot spot best. But when I said that we are trying to go for positive tumour localization, the reason is that we feel that what people want and need in the future, in addition to diagnosis, is treatment and they are never going to get treatment with an agent that doesn't concentrate in the tumour.

V.R. McCREADY: In this connection I would like to return to the discussion on computer processing when looking for hot areas. Computer processing tends to show up random "hot" areas which in fact are not significant. Normally one becomes used to a particular scanner with fixed settings and learns what is significant and what is not. When people use computers they often tend to change their settings each time, thus they are not comparing their examination against the normal pattern. This can produce false positive results.

H.J. GLENN: This discussion emphasizes how difficult it is to obtain meaningful data from the non-target area. In animal work we can get pretty good tumour uptake by isolating and counting the tumour; in regard to other tissue we can get separate tissue data very nicely (blood, skin and other areas), but the contribution that each of these make to the total non-target area is a problem that nobody basically has figured out. When we use the gamma camera approach, where we see the whole organ, we run into geometrical problems. For example, the left lobe of the liver is a lot thinner than the right one, and that is normally going to be lighter in the scan than the right lobe of the liver, and so one has to know what the contribution based upon distributional and geometrical factors is. These problems are extremely important in product development because these data have to be presented to regulatory agencies. Even after the radiopharmaceutical is in the clinical stage, when we try to get samples from autopsy from surgery, we don't know if the distribution has been influenced by the dying process, the anaesthesia, or the surgery itself.

IAEA-MG-50/14

FACTORS AFFECTING UPTAKE OF RADIOACTIVE AGENTS BY TUMOUR AND OTHER TISSUES*

R.L. HAYES
Medical Division,
Oak Ridge Associated Universities**,
Oak Ridge, Tennessee,
United States of America

Abstract

FACTORS AFFECTING UPTAKE OF RADIOACTIVE AGENTS BY TUMOUR AND OTHER TISSUES.
 The subject of factors that affect the uptake of radioactive agents by tumour and other tissues is a broad and complicated one. Factors that may have considerable bearing on the relative affinity of tumour tissue for such agents are, in general, the following: (1) vascularity, blood flow, and interstitial fluid space; (2) uptake of a protein-bound form of the agent; (3) cell proliferation; (4) presence of carrier; (5) capillary and cell permeability; (6) presence of inflammation; and (7) pH. Certain tumour-localizing processes are based on the specific properties of the tissues in question, i.e. radioiodine for functioning thyroid carcinoma, bone seekers for bone lesions, etc., but there is as yet still no clear indication of the basic process(es) involved in the tumour-localization of agents such as ^{67}Ga and others on which to base projections as to new materials that might be more specific for tumour tissue in general.

INTRODUCTION

Considerable progress has been made in recent years in the detection of tumours by radio-nuclide imaging. Nuclear medicine now has a battery of agents that, although not specific for tumour tissue alone, nevertheless do show considerable affinities for tumour tissues in general. We also do have some promising new agents for the detection of specific malignancies (i.e. 19-iodocholesterol for adrenal tumours and 7-iodochloroquine for melanomas). Bone malignancies and thyroid tumours and their metastases are, of course, special cases involving specific metabolic processes. But, if we are ever to have an agent for the specific detection of malignancies in general, it would appear that our best hope for such a development lies initially in an understanding of the similarities as well as the dissimilarities that exist between agents that have already been identified. Obviously the question as to why the radionuclides of gallium, indium, bismuth, cobalt, the higher-atomic-number rare earths, selenium (and others to perhaps a lesser extent) should, in rather simple forms, all show affinities for extracranial, non-osseous tumour tissue, although they are actually entirely different elements, is indeed an intriguing one. An examination of the similarities as well as the dissimilarities that these agents possess might well lead to the identification of other more specific agents (organic as well as inorganic).

I would like to discuss not only the similarities and dissimilarities involved in the uptake of various tumour-localizing agents in both normal and tumour tissue, but also the effect that various factors might have on the tissue distributions of tumour-localizing agents. These factors include the following: (1) vascularity, blood flow and interstitial fluids; (2) uptake of protein-bound substances; (3) cell proliferation; (4) effect of carrier; (5) capillary and cell permeability;

---

 * A portion of the work reported in this paper was supported by USPHS Research Grant CA 11858 from the National Cancer Institute.
 ** At time of presentation of paper, under contract with the United States Atomic Energy Commission.

(6) inflammation; and (7) pH. Certain other factors that affect tissue distribution, such as age, sex and dissociation constants, will also be treated. Since a major portion of our own work has been on studies of ^{67}Ga, comparisons will necessarily be heavily weighted in that direction.

VASCULARITY, BLOOD FLOW AND INTERSTITIAL FLUIDS

By whatever route substances are administered in vivo, their localization at sites other than the point of entry will usually involve transport to such sites by the blood. This necessarily raises the question of the degree of vascularity at such localization sites together with the rate of blood flow through the vasculature involved. The degree or rate at which the administered substance penetrates into the interstitial fluid and the size of the interstitial pool itself will also obviously be important factors in determining any transient as well as long-term localization of the substance at that site.

Because of the rapid growth frequently observed with tumours, it is not unreasonable to assume that the vasculature of such tumours might be greater than that observed with normal structures. Furthermore, in view of nutrient requirements for sustaining the increased growth rate, one might also expect the rate of blood flow to be enhanced. These effects may well occur although the presence of hypoxia in tumours might argue to the contrary [1].

Studies by Gullino [2] have indicated that blood volume and flow are reduced in transplanted rat and mouse tumours. He used transplantation and tissue isolation techniques which permitted him to make blood volume, blood flow, and interstitial fluid pool measurements in transplanted tumours that were served only by one incoming artery and one outgoing vein. He found that the vascular space of nine different rat and mouse tumours was, on the average, approximately 1/3 of that of liver. Furthermore, he also found that the blood flow through five different transplanted rat tumours (including the Novikoff hepatoma, a very rapidly growing tumour) was approximately 5% of that of the normal liver and ovary, the latter the host organ for the transplanted tumours. These results are contrary to the above expectations and serve to emphasize the fact that other factors might be exerting compensatory influences or that vascularity and blood flow are not nearly as important to increased tumour uptake as they might seem to be. These studies were of course carried out with transplanted tumours; nevertheless, there appears to be no compelling reason to assume that the blood volume and blood flow in spontaneous malignancies would be appreciably different. There are differences of opinion on this whole subject, of course [3, 4].

Gullino [2] also found, when he measured the interstitial fluid pool, that transplanted rat tumours had a considerably greater space than normal tissue, the tumours having approximately 2.5 times as great an interstitial space as the liver. Thus, assuming only normal and equilibrium transport of nutrients and other substances across the vascular membrane, tumour cells would be more than amply exposed for uptake. Regardless of whether permanent uptake is promoted on or in tumour cells, the increased size of the interstitial pool is very meaningful for transient concentration of radioactive materials in such tissues.

UPTAKE OF PROTEIN-BOUND SUBSTANCES

When a substance enters the vascular compartment, various degrees of binding of the substance can occur with the constituents of the plasma. This is particularly true with both ^{67}Ga and ^{111}In where binding to plasma proteins occurs almost immediately after intravenous administration, binding by ^{111}In being considerably stronger than that by ^{67}Ga. Protein-bound substances can subsequently enter the interstitial fluid spaces through pores in the microvasculature. Differences in the status of the microvasculature between tumours and the surrounding normal tissue may lead to transient increased concentrations of agents such as occurs in brain tumours.

Since preparations of ^{131}I-albumin had been previously observed to show some specificity for tumour tissue [5], presumably due to the rapidly growing tumour's ability to directly utilize intact proteins as well as amino acids for its increased nitrogen needs [6, 7], it was tempting, at the time that the uptake of ^{67}Ga by soft tissue tumours was first observed, to consider direct uptake of protein-bound ^{67}Ga as a possible major mechanism for the affinity of tumour tissue for ^{67}Ga [8]. This does not now seem likely to us [9]; the matter will be discussed at a later point.

Even though the extent of the binding of a tumour-localizing agent by plasma proteins may have little bearing on its ultimate long-term uptake by malignant tissues, the binding can have a decided effect on its temporal tissue distribution. The urinary excretion of an agent may be hampered by its binding to protein macromolecules, and the clearance of the agent from the blood will be slowed unless the tagged macromolecule(s) are removed intact by rapid tissue processes or by the dissociation of the macromolecular complex producing the free agent which is then in turn removed by the kidney or by uptake in tumour and various other tissues. Gallium-67, which is initially fairly strongly bound to plasma proteins, clears rather slowly from the blood and its tissue distribution only approaches completion after approximately one day. With ^{111}In, where plasma protein binding is stronger, the blood clearance is even slower. In the case of the higher-atomic-number rare earths, where the effective binding to plasma proteins is apparently low, tissue distribution is more or less complete within approximately 4 hours. These differences are illustrated by the comparative study shown in Fig.1. Although at an early time period (5 min) there is a surprisingly high apparent uptake of both ^{67}Ga and ^{171}Er in the tumour tissue on the left leg (possibly due to the relative mass of the tumour compared to that of adjacent normal tissue and also to the presence of a larger interstitial fluid pool), nevertheless it is apparent that the general clearance of ^{171}Er is much more rapid than that of ^{67}Ga, with the ^{171}Er concentration in the tumour peaking at approximately 4 hours. With ^{67}Ga the tumour is still accumulating the radionuclide to some extent even at 18 hours.

CELL PROLIFERATION

The rapid growth rate observed with certain malignancies, where tagged metabolites would be expected to be taken up more rapidly and to a greater extent than in normal tissues, may be a promising basis for radiopharmaceutical development, particularly with ^{11}C-labelled agents, but at present, with the exception of the limited success achieved with radioiodine-labelled proteins and ^{75}Se-labelled selenomethionine, no effective agents or techniques have been forthcoming using this particular approach. In our particular studies of ^{67}Ga in various transplanted tumour models we have seen no enhanced gross tumour uptake of ^{67}Ga in the presence of increased growth rates, but instead have found that slow-growing tumours have better uptakes, possibly because there is less tendency toward necrosis which would lower the overall ^{67}Ga tumour concentration [10]. Hepatectomy in normal rats also appears to have only a minor effect and then only at an early time period in the regeneration process (R.L. Hayes and B.L. Byrd, unpublished results). The work of Otten et al. [11] has also shown that ^{3}H-thymidine is taken up to a much greater extent in intestinal crypt cells than is ^{67}Ga.

EFFECT OF CARRIER

From an historical standpoint gallium radionuclides afford an excellent example of the effect that the presence of carrier can have on the uptake of a tumour-localizing agent in both normal and malignant tissues. Undoubtedly the inadvertent presence of stable gallium in preparations of

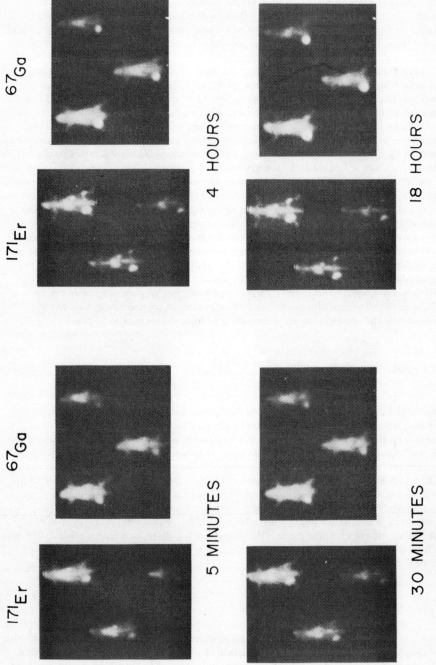

FIG.1. Serial scans of ^{171}Er and ^{67}Ga-citrate in Buffalo rats bearing Morris 5123C transplanted hepatomas. A total of 100 000 counts was collected in each exposure.

^{72}Ga was responsible for the fact that early workers did not observe a concentration of this radionuclide in soft tissue tumours [10]. Simultaneous administration of stable gallium with ^{67}Ga will greatly increase blood clearance, urinary excretion, and to some extent bone deposition, while the concentration in other normal tissues (except for the kidney) will be greatly reduced. Stable gallium administration also reduces radionuclide concentration in non-osseous tumours. Although carrier gallium does not change the relative distribution of ^{67}Ga in the subcellular organelles, it does affect its macromolecular association drastically, even when administered at very low levels [12]. Presumably because of its close chemical similarity, stable indium produces an effect similar to that of stable gallium on ^{67}Ga distribution in both normal and malignant tissues. Both stable indium and gallium in turn have similar effects on the distribution of carrier-free indium radionuclides [13]. Interestingly enough, the tissue distributions of bismuth and the higher atomic number rare earths are also subject to carrier effects [14].

Similarly with the radionuclide ^{206}Bi, simultaneous administration of stable bismuth in general decreases the ^{206}Bi concentrations in most tissues including that in transplanted tumour. With the rare earth radionuclides at $\sim$ 1 mg/kg of carrier there is, compared with the carrier-free material, a pronounced decrease in concentration in tumour tissue and bone, and a corresponding extreme elevation in the deposition in reticuloendothelial tissues. At levels of $\sim$ 10 μg/kg, where there was no change produced in rare earth tumour concentration, the increased deposition in reticuloendothelial tissue was still quite marked.

These observations emphasize the importance of carrier effects where tumour-localizing radiopharmaceuticals are involved and in turn suggest that reinvestigation of previously reported distribution studies done only with low specific activity preparations might well be productive.

CAPILLARY AND CELL PERMEABILITY

We can distinguish two general types of permeability that may relate to the transient and long-term uptake of radiopharmaceutical agents by normal and malignant tissue. The first involves the permeability of the microvasculature with resulting entry of macromolecules into the interstitial space. The second involves the permeability of the plasma membrane of the cell controlling entry into the cell proper. (There could, of course, be direct binding of the agent or labelled macromolecule to the plasma membrane itself.) Potchen et al. [3] have speculated on the importance of differences in the microvasculature as it relates to entry of protein-bound substances into the extravascular compartment. Even though this type of permeability may be a major factor in the transient concentration of various agents in lesions such as brain tumours, the permeability of the plasma membrane to the unbound agent may be even more important in long-term retention of agents in tumour tissue.

This is suggested by our observations in animals on alterations in the uptake of ^{67}Ga in normal and malignant tissues that are produced by administration of stable gallium and scandium [10]. Administration of stable gallium with ^{67}Ga, as previously mentioned, produces greatly decreased uptakes of ^{67}Ga in normal soft tissues and tumour together with a dramatic increase in urinary excretion and an enhanced blood clearance. Plasma protein binding sites are saturated by stable gallium and the excess gallium together with the ^{67}Ga label is then rapidly excreted [15]. The element scandium also blocks ^{67}Ga plasma binding and produces the same effect on ^{67}Ga excretion and normal soft tissue uptake but, contrary to the effect of stable gallium, which saturates gallium tumour-binding sites, the uptake of ^{67}Ga in tumour tissue either remains the same or is somewhat increased [10]. These results have led us to speculate that the major pathway of ^{67}Ga into the intracellular space in tumour tissue is via dissociation of the protein-^{67}Ga complex(es) to yield an "unbound" agent, possibly inorganic, to which the tumour plasma membrane is "hyperpermeable" and that pinocytosis of the protein-^{67}Ga complex is a minor process with tumour tissues, whereas it may be a major one with certain soft tissues such as liver [9]. If pinocytosis of protein-bound

TABLE I. DISTRIBUTION OF VARIOUS TUMOUR-LOCALIZING RADIONUCLIDES IN BUFFALO RATS BEARING *Staphylococcus aureus* INDUCED ABSCESSES[a]

	Radionuclides					
	^{67}Ga	^{111}In	^{169}Yb[b]	^{203}Hg	^{206}Bi	^{75}Se
Concn (%/g) in abscess wall	3.4	2.9	1.3	1.3	0.3	0.6
Concn (%/g) in necrotic fluid	0.5	0.6	0.6	0.5	0.6	0.7
Ratio of abscess concn to normal:						
Liver	4.0	2.3	2.9	1.9	0.5	0.6
Spleen	2.6	1.6	2.0	1.6	0.6	0.8
Kidney	5.2	1.0	0.9	0.1	0.03	0.2
Lung	11.0	4.8	7.0	2.3	1.40	1.0
Muscle	15.0	17.0	40.0	19.0	60.0	6.4
Femur	2.5	3.8	0.4	4.7	1.4	2.5
Blood	28.0	7.5	58.0	8.6	36.0	1.0

[a] Four or more animals per group; 24-h distribution.
[b] Separate group. Other radionuclides studied on groups of animals inoculated at same time.

^{67}Ga is a major pathway in the uptake of ^{67}Ga by liver, then blocking of the protein binding should result in a decreased ^{67}Ga uptake. This indeed occurs although increased bone deposition and urinary excretion might contribute to the decrease as competing processes. With administration of scandium, where the plasma ^{67}Ga-binding sites are also blocked, the uptake of ^{67}Ga in tumour is not only as high as it is in the normal protein-bound form, but it is at the same time necessarily competing with increased bone deposition and urinary excretion. This fact suggests that even in the absence of a blocking of protein-binding sites for ^{67}Ga, the major pathway into tumour tissue is possibly through an unbound form rather than by pinocytosis of protein-bound ^{67}Ga. We are of course dealing here with only one agent, but these results do suggest that a pathway similar to this one proposed for ^{67}Ga might also be involved with the rare earths and more probably with indium, since the intracellular associations are similar (see later).

INFLAMMATION

Inflammatory lesions are now well known to have an affinity for ^{67}Ga. In fact, recent reports indicate that, although it was at first considered only an interference with tumour detection techniques, this affinity may have diagnostic benefit in its own right [16, 17]. After noting in our early work that ^{67}Ga did tend to localize in animal abscesses, we have made it a habit to test for localization of this sort with other agents. We have found that ^{111}In and the rare earth radionuclide ^{169}Yb [14] show appreciable concentrations in inflammatory lesions. Table I shows a recent comparative study we made of the uptake of various tumour-localizing agents in an experimental animal abscess. Gallium-67 had the highest abscess uptake of those tested, with ^{111}In, ^{169}Yb, and ^{203}Hg also showing some affinity. These results suggest that rare earth radionuclides might merit further investigation as agents for the detection of abscesses.

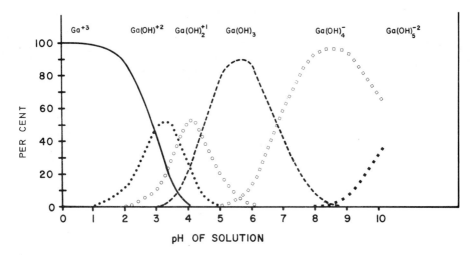

FIG.2. Species relationships for gallium at various pH values calculated from selected hydrolysis constants [22, 23].

An interesting point is that these different tumour-localizing agents share to some degree an affinity for inflammatory lesions. The mechanisms of these affinities are not known in either tumours or inflammatory lesions although in the case of ^{67}Ga the use of ^{67}Ga-labelled white cells for abscess detection [18] and the reported observation of such cells in abscess walls [16] would indicate the involvement of a cellular response to infection as a possible basis for the uptake in inflammatory processes. The uptake of ^{67}Ga in the lysosomes of mononuclear phagocytic cells has been well demonstrated by Swartzendruber et al. [19]. Even so, it is tempting to speculate that there may be some common immunological basis for concentrations in the two different types of lesion.

EFFECT OF pH

The existence of a difference in the hydrogen ion concentration in the milieu of tumours and normal tissues has frequently been suggested as being the basis for differences in behaviour between these two types of tissue. Gullino et al. [20] have reported measurements in isolated transplanted tumours that indicate that the interstitial fluid in tumours may be as much as 0.3–0.4 of a pH unit lower than it is in normal tissue. It is certainly not unreasonable to postulate that such a difference in pH might predispose tumour-localizing agents toward an increased uptake in malignancies. Such an effect has not as yet been demonstrated, although Glickson et al. [21] have reported an increase in in-vitro ^{67}Ga labelling of L1210 cells when the pH was decreased.

An interesting fact that does stand out about those tumour-localizing agents that show the highest degree of affinity for tumour tissue (i.e. ^{67}Ga, ^{111}In, ^{206}Bi, and the rare earths) is that they are generally subject to hydrolysis and can exist in multiple hydroxo states at or near physiological pH. Figure 2 illustrates possible species relationships for gallium at various pH values based on the use of selected hydrolysis constants [22, 23]. (The literature contains widely varying values for such constants and, therefore, this plot is to be considered only approximate.) Figure 2 predicts that at physiological pH $Ga(OH)_3$ and $Ga(OH)_4^-$ forms will predominate. This in turn raises the interesting point that $Ga(OH)_3$ (monomer or polymer) being neutral would not be subject to electrostatic effects.

TABLE II. EFFECT OF AGE ON THE TISSUE DISTRIBUTION (24 h) OF ^{170}Tm IN THE MALE BUFFALO RAT[a]

Tissue	Per cent of dose/g[b]		Significance (P)
	Administered to 2-month-old rats	Administered to 5-month-old rats	
Liver	0.36	0.48	<0.001
Spleen	0.19	0.29	<0.001
Kidney	0.76	2.10	<0.001
Lung	0.14	0.29	<0.001
Muscle	0.02	0.04	<0.001
Femur	4.60	4.30	<0.05 >0.02
Marrow	0.26	0.38	<0.01 >0.001
Blood	0.01	0.02	<0.001
Average weight	190	370	

[a] Groups of five animals each.
[b] Per cent administered dose/g normalized to body weight of 250 g.

The existence of a lower pH in tumour tissue would also favour the $Ga(OH)_3$ species, suggesting that pH differences between tumour and normal tissues may be involved to some extent in the increased uptake of ^{67}Ga by malignancies. On the other hand calculations indicate that indium at a similar pH should be present in the hydroxide form to a greater extent than is gallium, and yet in our hands ^{111}In has shown lower uptakes in animal tumours than ^{67}Ga [10]. Nevertheless, pH effects, particularly as they might relate to hydrolytic effects, may be of importance for cationic-type tumour-localizing agents.

OTHER FACTORS

Age and sex

Rather dramatic effects are observed in the tissue distribution of ^{67}Ga in rats as the result of differences in sex and age, the major reticuloendothelial tissues and muscle showing highly significant differences in ^{67}Ga uptake [9]. Interestingly enough rat tissue distribution of the higher atomic number rare earths is also affected by age and sex although the differences seen are not nearly as pronounced (Tables II and III). Possibly indium radionuclides will also show age and sex effects, since indium and gallium are chemically so similar.

In an extension of the initial observations on differences in ^{67}Ga tissue distribution between male and female rats, we have studied the effect of castration and administration of sex hormones (R.L. Hayes and B.L. Byrd, unpublished results). Castration of male rats produced ^{67}Ga tissue distributions resembling those of the female. Administration of testosterone to castrated males produced tissue distributions approaching those in normal males. Administration of testosterone to females also resulted in distributions approaching that of normal males, while administration of oestradiol to normal males altered their tissue distribution pattern toward that of the female.

TABLE III. TISSUE DISTRIBUTION (24 h) OF ^{170}Tm IN THE MALE AND FEMALE FISCHER RAT[a]

Tissue	Per cent of dose/g[b] in:		Significance
	Male rats	Female rats	(P)
Liver	0.74	0.56	0.001–0.01
Spleen	0.57	0.31	<0.001
Kidney	0.62	0.52	0.02
Muscle	0.03	0.02	<0.001
Femur	5.00	4.30	<0.001
Marrow	0.33	0.22	—
Blood	0.02	0.01	<0.001

[a] Groups of five animals each.
[b] Per cent administered dose/g normalized to body weight of 250 g.

TABLE IV. COMPARISON OF THE SUBCELLULAR DISTRIBUTION OF ^{111}In, ^{167}Tm, AND ^{206}Bi WITH THAT OF ^{67}Ga IN THE MORRIS 5123C HEPATOMA

	Fractions					
	I[a]	II[b]	III[b]	IV$_p^b$	IV$_i^b$	IV$_s^b$
^{67}Ga	2.1	3.8	27.6	53.3	5.6	9.7
^{111}In	5.0	5.0	23.0	53.6	5.5	12.9
^{167}Tm	7.9	7.0	16.8	54.3	4.7	17.1
^{206}Bi	11.4	25.2	16.5	31.1	8.3	18.9

[a] Per cent of total counts put in SZ-14 zonal rotor.
[b] Per cent of total counts put in B-29 rotor for SPR zonal centrifugation procedure.

Subcellular distribution

Considerable evidence now exists to indicate that the main intracellular deposition site for ^{67}Ga in normal and tumour tissue is the lysosome [19, 24]. Dr. D.H. Brown, at our institution, has recently isolated by zonal ultracentrifuge techniques still another class of ^{67}Ga-binding particles that is considerably smaller than normal lysosomes (D.H. Brown et al., unpublished results). These particles are quite prominent in homogenates of hepatoma and less so in normal liver. The exact identity of the particles has not yet been established, but the fact that they are associated with high levels of acid phosphatase suggests that they may be lysosomal in nature.

Dr. Brown has also compared the subcellular distribution of ^{67}Ga with that of ^{111}In, ^{167}Tm and ^{206}Bi. The initial results of this study are shown in Table IV. Subcellular fractionation was carried out using a B-XXIX zonal rotor and a rate-zonal sequential-product-recovery (SPR) procedure devised by Brown [24]. Fraction I in Table IV consisted mostly of nuclei, fraction II mostly of mitochondria, fraction III of normal lysosomes, and fraction IV$_p$ of the new small particle population.

TABLE V. RELATIVE CONCENTRATION OF ^{58}Co-CITRATE AND BLEOMYCIN IN RAT AND MOUSE TRANSPLANTED TUMOURS[a]

	5123C Hepatoma		P-1798 Lymphosarcoma	
	Citrate	Bleomycin	Citrate	Bleomycin
Tumour concn (%/g)	0.3	0.05	3.7	1.7
Ratio of tumour concn to normal:				
Liver	0.5	0.2	0.6	0.8
Spleen	2.9	0.6	5.6	2.7
Kidney	0.3	0.1	1.2	1.3
Lung	2.0	2.0	1.9	4.1
Muscle	16.0	27.0	20.0	26.0
Blood	5.7	12.0	5.6	20.0
Excretion (%)	79	88	53	78

[a] Three animals per group.

Fractions IV$_s$ and IV$_i$ were the supernatant soluble zone and the zone intermediate between IV$_p$ particles and IV$_s$, respectively. It is apparent that the subcellular distributions of ^{111}In and ^{167}Tm are very similar to that of ^{67}Ga suggesting that similar subcellular particles might be involved in the intracellular associations of ^{111}In and ^{167}Tm. The distribution of ^{206}Bi on the other hand was rather diffuse. In this connection it is of interest that ^{67}Ga and ^{111}In showed similar associations with a 40 000-mol. wt macromolecular component present in aqueous extracts of tumour tissue [12], whereas ^{206}Bi showed little association with this component.

Anions and chelates

In general, studies with various anionic agents have indicated that the particular anionic form used with ^{67}Ga does not appreciably affect its distribution in normal and tumour tissue so long as the complexing or chelating agent used does not bind the ^{67}Ga too strongly. The actual amount of agent administered also does not appear to be important [10]. Apparently this arises from the fact that ^{67}Ga is very rapidly bound to plasma proteins and, parenthetically, there is thus actually no reason, if desired, why ^{67}Ga should not be administered as the chloride, so long as the ^{67}Ga injection solution is sufficiently acid to avoid colloid formation. It would seem reasonable to assume that much the same behaviour would occur with other simple metals like gallium.

Bleomycin, a cancer chemotherapeutic agent, has recently received considerable attention as a "chelate carrier" for, in a sense, converting various radionuclides into tumour-localizing agents. Cobalt-57-bleomycin has especially received attention in this regard. A brief study we carried out with ^{58}Co-bleomycin, however, raises a question in our minds as to whether bleomycin is actually any more than a modest enhancer of the actual nature of cobalt itself as a tumour-localizing agent. Table V shows that ^{58}Co-citrate had practically as good a distribution as did the bleomycin form, both in terms of absolute tumour uptake and tumour-to-non-tumour ratios.

As indicated before, one would normally expect that those chelating agents for a particular metal that form chelates that have high stabilities would produce a masking of the nature of the metal and that the metal chelate would be rapidly excreted. Such is the case with ^{67}Ga-EDTA and ^{67}Ga-HEDTA. However, we have recently observed that with DTPA, which is a much stronger

TABLE VI. EFFECT OF DTPA AND EDTA ON THE TISSUE DISTRIBUTION (24 h) OF ^{67}Ga IN THE FISCHER 344 RAT-BEARING RFT TUMOURS[a]

Tissue	Citrate	EDTA	DTPA
		(Per cent administered dose/g)	
Liver	1.50	0.045	0.90
Spleen	2.80	0.130	1.30
Kidney	0.81	0.089	0.61
Lung	0.30	0.013	0.21
Muscle	0.17	0.001	0.12
Femur	1.30	0.038	0.87
Marrow	1.70	0.039	0.78
Blood	0.18	0.007	0.15
Tumour	2.00	0.055	1.70

[a] Groups of four male animals each with data normalized to body weight of 250 g.

chelator of gallium than is EDTA, the tissue distribution of ^{67}Ga is surprisingly similar to that of citrate (and in some respects superior). Table VI shows the results of a comparison study of the distributions of ^{67}Ga-citrate, ^{67}Ga-EDTA and ^{67}Ga-DTPA.

We feel that the basis for this anomalous behaviour of ^{67}Ga-DTPA probably lies in the fact that although the stability of the gallium-DTPA chelate is much greater than that of gallium-EDTA, the rate of dissociation of gallium-DTPA is much higher than that of gallium-EDTA. In fact the rate of exchange of gallium ions with the gallium in gallium-EDTA is reported to be practically nil at physiological pH [25]. No data are available on the rate for gallium-DTPA dissociation.

This behaviour of gallium-DTPA does, however, indicate that the equilibrium value for the dissociation constant of a complex or chelate is not necessarily a valid measure of its behaviour in vivo. Further, it would appear that perhaps chelating agents that are not now thought of as being useful (because of their stability constants) may indeed be quite stable in vivo for appreciable periods of time where dissociation rates are extremely slow once the chelate has been formed.

ACKNOWLEDGEMENTS

The author is happy to acknowledge the help and advice of Drs. G.A. Andrews and C.L. Edwards (Medical Division, Oak Ridge Associated Universities) in the preparation of this manuscript.

REFERENCES

[1] TANNOCK, I.F., The relation between cell proliferation and the vascular system in a transplanted mouse mammary tumour, Br. J. Cancer 22 (1968) 258.
[2] GULLINO, P.M., The internal milieu of tumors, Prog. Exp. Tumor Res. 8 (1966) 1.
[3] POTCHEN, E.J., et al., Pathophysiologic basis of soft tissue tumor scanning, J. Surg. Oncol. 3 6 (1971) 593.
[4] REINHOLD, H.S., The relationship between tumour vascularisation and response to radiotherapy, TNO Nieuws 27 (1972) 737.

[5] BONTE, F.J., et al., Radioisotope scanning of tumors, Am. J. Roentgenol., Radium Ther. Nucl. Med. **100** (1967) 801.
[6] BABSON, A.L., WINNICK, T., Protein transfer in tumor-bearing rats, Cancer Res. **14** 8 (1954) 606.
[7] GHOSE, T., et al., Uptake of proteins by malignant cells, Nature (London) **196** (1962) 1108.
[8] EDWARDS, C.L., HAYES, R.L., Scanning malignant neoplasms with gallium-67, J. Am. Med. Assoc. **212** (1970) 1182.
[9] HAYES, R.L., BROWN, D.H., "Biokinetics of radiogallium", Nuklearmedizin (Proc. 12th Int. Meeting of the Society of Nuclear Medicine), F.K. Schattauer-Verlag, Stuttgart-New York (1975) 837.
[10] HAYES, R.L., EDWARDS, C.L., "New applications of tumour-localizing radiopharmaceuticals", Medical Radioisotope Scintigraphy 1972 (Proc. Symp. Monte Carlo, 1972) **2**, IAEA, Vienna (1973) 531.
[11] OTTEN, J.A., et al., Localization of gallium-67 during embryogenesis, Proc. Soc. Exp. Biol. Med. **142** (1973) 92.
[12] HAYES, R.L., CARLTON, J.E., A study of the macromolecular binding of ^{67}Ga in normal and malignant animal tissues, Cancer Res. **33** (1973) 3265.
[13] HAYES, R.L., et al., ^{113m}In as a possible bone scanning agent, J. Nucl. Med. **9** (1968) 323.
[14] HAYES, R.L., et al., A comparison of the tissue distribution of ^{67}Ga and the rare earth radionuclides, J. Nucl. Med. **15** (1974) 501.
[15] HARTMAN, R.E., HAYES, R.L., The binding of gallium by blood serum, J. Pharmacol. Exp. Ther. **168** (1969) 193.
[16] LITTENBERG, R.L., et al., Gallium-67 for localization of septic lesions, Ann. Intern. Med. **79** (1973) 403.
[17] FRATKIN, M.J., et al., Gallium scanning and inflammatory lesions, Ann. Intern. Med. **80** 1 (1974) 114.
[18] BURLESON, R.L., et al., Scintigraphic demonstration of experimental abscesses with intravenous ^{67}Ga citrate and ^{67}Ga labeled blood leukocytes, Ann. Surg. **178** (1973) 446.
[19] SWARTZENDRUBER, D.C., et al., Gallium-67 localization in lysosomal-like granules of leukemic and non-leukemic murine tissues, J. Natl. Cancer Inst. **46** (1971) 941.
[20] GULLINO, P.M., et al., Modifications of the acid-base status of the internal milieu of tumors, J. Natl. Cancer Inst. **34** (1965) 857.
[21] GLICKSON, J.D., et al., Effects of buffers and pH on in vitro binding of ^{67}Ga by L1210 leukemic cells, Cancer Res. **34** (1974) 2957.
[22] NAZARENKO, V.A., et al., Determination of the hydrolysis constants of gallium ions, Russ. J. Inorg. Chem. (English Transl.) **13** 6 (1968) 825.
[23] SHEKA, I.A., et al., The Chemistry of Gallium, Elsevier Publishing Co., Inc., New York (1966) 45.
[24] BROWN, D.H., et al., The isolation and characterization of gallium-binding granules from soft tissue tumors, Cancer Res. **33** (1973) 2063.
[25] SAITO, K., TSUCHIMOTO, M., A kinetic study of the isotopic exchange of gallium between the gallium ion and gallium ethylenediamine-N,N,N', N'-tetraacetate and N-2-hydroxyethyl ethylenediamine-N,N', N'-triacetate in water, J. Inorg. Nucl. Chem. **23** (1961) 71.

DISCUSSION

V.R. McCREADY: In a recent paper in the Journal of Nuclear Medicine on ^{67}Ga uptake in hepatectomized rats no increase in uptake could be found following hepatectomy. The results were taken as a reflection of the lack of correlation between ^{67}Ga concentration and growth rate. Taylor and Hammersley, too, have been studying it but on a longer time scale and they got the lowest uptake of gallium in their system 18 hours after hepatectomy. They stress that you can't just use a single measurement to state that there is no correlation between growth rate, regeneration and gallium uptake. We have also found differences between the protein binding of ^{67}Ga in vitro and in vivo. One has to be careful when comparing different studies on gallium uptake since the conditions seem to vary from one experiment to another. We must carry out more basic work before we move on to the more complicated aspects of the gallium studies.

R.L. HAYES: The results mentioned in my paper originated from work we did some time ago and I made the parenthetical remark that we didn't see any pronounced effect. The effect we did observe was in the early period (around 12 to 18 hours) but again it was not particularly dramatic. As to the comments that you raised about the effect of citrate, may I ask you if you were dealing with an intact system or a tissue culture?

V.R. McCREADY: We did some simple experiments trying to label proteins in vitro to see whether temperature and the level of citrate were significant. Certainly in in-vitro experiments temperature had an effect on protein labelling. Various concentrations of citrate greatly affect the excretion of gallium in animals, and indirectly the amount of gallium taken up in tumours.

R.L. HAYES: In our studies, which were carried out by Hartman and reported in 1969, the amount of citrate had very little effect in equilibrium dialysis experiments when gallium was added to the plasma proteins until we got up to very high citrate levels. In our in-vivo work in animals and in whole-body measurements on patients, we found no citrate effect running all the way from a level of 7 mg/kg down to 0.01 mg/kg, in other words at a level where you are approaching practically no citrate. When you talk about protein studies were you using plasma proteins or extracts from tissues?

V.R. McCREADY: Plasma proteins.

R.L. HAYES: Obviously, you have a competing process.

V.R. McCREADY: You mention that there was more uptake and retention in male rats. Do you think this is related to the fact that there is little uptake of gallium in breast tumours in humans? Can you see any relation?

R.L. HAYES: No, I really can't.

W.H. BEIERWALTES: As to vascularity and blood flow in tumours, considerable work appears right now from different institutions about the "increased blood flow effect" in causing increased uptake of technetium polyphosphate and diphosphonate in bone scintigraphy. It seems as if there are at least two mechanisms of uptake of technetium diphosphonate. The first one from increased blood flow, which seems to result in more of the technetium diphosphonate going to the crystal in an ion-exchange reaction. There may be a marked increase in the uptake of the bone over a considerable area proximal and sometimes distal to the tumour. This does not relate to blood volume because they don't get the same picture if they use ^{51}Cr red cells or ^{131}I human serum albumin for their studies. Then in a later phase of the uptake you have actually more chemical incorporation of the diphosphonate, for example into the bone crystals as such. But it is quite striking that you have a marked increase in uptake several inches either side of the tumour.

A second point on blood flow is that there could be a very good blood flow, either normal or even increased, in a solid tumour, where the active proliferating cells are. All solid tumours, at least at some stage, have an area of necrosis. If you look at the tumours microscopically, you will find that there are tiny vessels and lots of growing cells around the vessels, and often necrosis outside this area. If you measure blood flow to the entire tumour, including the necrotic areas which are not always visible, you might not find an increased blood flow.

Secondly, you might have oedema with decreased vascularity in those areas. A couple of days ago, our pathologist demonstrated an astrocytoma grade 4 with rather massive oedema in some areas and other areas with normal vessels and actually haemorrhagic necrosis, a picture varying between good vascularity and areas with massive oedema.

The next point I want to stress is permeability. In the past few weeks we have studied some new iodochloroquines and we wondered why this much larger molecule showed a greater uptake than our original compounds. We were interested in some work published by Hansch who had demonstrated in studies on barbiturates that the octanyl:water partition coefficient is a more reliable guide for their effectiveness than all other parameters. He demonstrated that the barbiturates which were less soluble in water had a better uptake in the tissues, presumably as a result of cell membrane transport. Following this line we used chloroform versus a buffer that had the composition of extracellular fluid with our new chloroquine analogues. We achieved a perfect correlation of uptake using five different compounds. The only exception was a demethylated compound which didn't fit into the scheme. Now, in this connection, I just wonder if you have had the opportunity to study the octanyl:water partition coefficient in relation to gallium, the way you have been using it, in various complexes?

Finally, there is considerable evidence in the literature that in studying tumour uptake one should also study the uptake in the wall of the blood vessels leading to the tumour. Boron has been found to adhere strikingly to the intima of the arteries leading into the brain tumours and also radio-iodinated antibodies against gliomas have been shown to have a high uptake in the walls of vessels leading to the tumours, but not a high concentration in the tumours themselves. Lastly, I have been intrigued recently by the uptake in mitochondria and I wonder if you could tell me what time interval you used in the study of gallium and bismuth in the mitochondrial fraction.

R.L. HAYES: I agree with your comments on vascularity and permeability. I was raising, in the case of animal data, the question of how valid it is to try to make generalizations. About gallium I can say that it appears to be predominantly neutral and in that sense nonpolar at physiological pH. Whether this has any significance or not I do not know. As to the question about deposition of gallium in vessel walls, I have no observations, but these questions could be answered by autoradiography. In the case of bismuth, subcellular distributions were measured at 24 hours. We have looked at the distribution of gallium at early time periods. Surprisingly enough within five minutes to half an hour the ^{67}Ga activity present in the cell is found in a bound form and is tied up predominantly to the two particle fractions I was talking about. So, when it passes the plasma membrane, by whatever process, it is very rapidly incorporated into organelles and does not remain in the cell sap for any appreciable time.

W.H. BEIERWALTES: You have studied gallium uptake when you studied the subcellular fractions and found it primarily in the large and the small lysosomes at 15 to 30 min. Do you recall the mitochondrial concentration at that time?

R.L. HAYES: By the sequential product recovery technique I was talking about, you also obtain mitochondria as a part of the separation process. The activity in the mitochondria was very slight. As you probably noticed, this is not so in the case of bismuth.

D. COMAR: When we speak about carrier-free elements, which we want to use for tumour localization, does it mean that the amount to be injected into the patients must be lower than the amount in, for instance, the plasma pool? With gallium the amount of carrier must be very small if you want a high uptake in a tumour. If you use copper, which is taken up by the tumour just as cobalt, then the amount of copper can be higher than the amount which you use, if you inject gallium. Is it only a question of pool size, or does it depend on other conditions?

R.L. HAYES: I think you have to consider two processes here. One certainly is the plasma pool. In the case of copper the total pool might very well not be so much the amount of copper that is bound to ceruloplasmin but what is actually present in an equilibrium state with other tissues. The answer to this sort of thing can only be obtained by experiment. Copper being a normal constituent of the biological system, you obviously have the effect of its pool size. However, this doesn't mean that you couldn't overload the pool.

In the case of gallium there is no evidence to indicate that it is a requirement or normal constituent, it is just a trace impurity in tissue. But you still have to distinguish between several pools. One would be the plasma-binding pool. Once you saturate that you get a rather dramatic effect on the excretion and on the deposition in bone, and presumably because of the competitive urinary excretion you get a decreased deposition in normal tissues. Then there is another important pool in tissues where ultimate binding of gallium occurs. I think that it is mainly macromolecularly bound in tumour tissue, whether it is associated with particles or not.

How much or what the capacity of that binding is, is a different thing. Initially it looked like about one-tenth of a milligram per kilogram of administered stable gallium was required before the deposition in normal tissues as well as in tumour was affected. But when we looked at the association of gallium with the macromolecular species present in extracts of tumour tissue we found that as little as 10 to 25 µg of stable gallium administered per kilogram of body weight would almost completely block this binding which seems to be characteristic of tumour tissue.

H.J. GLENN: Dr. Hayes was talking about plasma-protein binding in general terms. I wonder whether we won't have to come down to considering specifically the proteins that bind, for instance, gallium or indium, such as transferrin, and that may make quite a difference because we may come into a situation where the iron concentration plays an important role. If we consider a binding to transferrin rather than a binding to globulin or albumin, iron in the circulation will enter into the equilibrium and may tend to free some of the tracer compounds from the transferrin. Perhaps we should look a little more into this by using bi-functional linking compounds that bind one end to a specific protein and also bind at the other end to different radionuclides.

In addition, I would like to comment on specific activity. We were interested in ^{169}Yb-citrate as a tumour-localizing agent, but we couldn't use some ^{169}Yb because the reactor facilities gave a compound of low specific activity; we found 50–60% deposited in the bones.

With reference to the membrane permeability, I want to mention that we studied gallium uptake in our melanoma cell line and the influence of scandium on this uptake. We found that small amounts of scandium did increase the uptake until we reached a level at which we obviously had a cell toxic effect.

When we compared in mice the liver uptake of gallium chloride, as opposed to gallium citrate or gallium lactate, we found big differences in the liver concentration within the first two hours. With the chloride, we found an early peak in the liver uptake and then it fell off again, whereas the citrate and lactate concentrations were perfectly level. We have speculated that this might indicate some natural process in the liver, which metabolizes citrate and lactate, before we get the material into the liver cells, and that this would account for the constant liver levels which could indicate a constant rate of metabolism of the citrate and lactate.

Furthermore, pH is very crucial. With gallium you have to work at a pH of 3 or below to get the distribution pattern of the gallium ion. If the pH is above 3, it is in a colloid form and will largely end up in the liver. With ^{111}In-chloride, we have to work at a pH of about 1.5; we get completely incomprehensible data if we work at pH 3. Even if you can't see any precipitated material with carrier-free substances you may have a dispersed colloid and not a true solution.

Speaking of chelating agents, I would like to comment on our work with gallium-iron-DTPA. Originally Wagner and others added a few micrograms of iron to the preparation because – being afraid of carrier-free chemistry abnormalities – they thought that it might be better to add a few micrograms of iron as carrier. All our work was also done with a few micrograms of iron in the preparation. When we then studied the uptake in the melanoma cell of gallium we observed that under the conditions of keeping the iron below or well within the chelating capabilities of the DTPH, the uptake within the cell increased, if we added more iron. So perhaps we should pay more attention to the addition of "carriers" to the carrier-free state. This effect cannot be related to transferrin or other binders in this particular cell system.

W.H. BEIERWALTES: In the French literature, you can find articles by Buffe and her co-workers on an α_2H globulin which some people think is an extremely specific cancer antigen. Last month an article was published in Archives of Biochemistry and Biophysics on a derivative of Buffe's α_2H globulin found in breast cancer but not in the normal breast. The compound has been characterized as a ferritin and Buffe and her co-workers are now beginning to identify a whole series of ferritins which might very well be related to your observation of iron in melanomas. It is possible that each tumour has a unique ferritin and a unique concentration of iron which might bear on the uptake of certain radionuclides or radionuclide compounds in tumours.

K. HISADA: Dr. Hayes' slide with the ^{206}Bi data reminds me of a very old paper on the tumour affinity of bismuth in Strahlentherapie, 1930, Vol. 37, page 751. The author, Herbert Kahn, discovered that natural radium, which is identical with ^{210}Bi, can accumulate in tumours, and he mentioned that if adequate equipment for beta radiation measurements were available, ^{210}Bi could be used for "photographic" diagnosis of a primary tumour and its metastatic lesions. It's surprising that such a study on the ^{210}Bi affinity for tumour tissues was carried out 40 years ago without any artificial radionuclides and without Geiger-Müller counters. The reason is that heavy metals were used for incurable diseases, for instance lead for skin disease, gold for tuberculosis and bismuth for cancer. Their body distribution and their toxicity were the main object of the studies at that time. Except for Kahn's wrong remark on the beta emission, had he described gamma-ray equipment instead of beta-ray equipment, his remarks might very well be valid today.

R.L. HAYES: The reason for our studies on bismuth is that it has a relative tumour uptake that is equivalent to, if not better than, ^{67}Ga. The drawback is that bismuth does not have any radionuclides which are useful. ^{206}Bi is far too high in energy and ^{204}Bi has a very short half-life, and would be expensive to produce.

D. COMAR: Which isotope of bismuth did you use and which chemical form did it have?

R.L. HAYES: I used ^{206}Bi. It was carrier-free in the citrate form (1 mg of citrate/kg and adjusted to pH 7).

K. HISADA: ^{210}Bi emits alpha particles in addition to beta and gamma rays. Kahn's paper also includes figures on the relative distribution in the human body.

H.J. GLENN: In our experience it has been almost impossible to solubilize preformed bismuth citrate by itself; it is a very insoluble compound. But if you add enough sodium citrate to bring it into solution, this probably involves a sodium bismo-citrate type of molecule. I think this would probably move the bismuth into an anionic form.

R.L. HAYES: It was our presumption that we were dealing with Bi^{+++} and that we were simply forming a citrate complex of some sort. We didn't at that time carry out any ultrafiltration experiments to determine whether the preparation was colloidal, but the very fact that we did not find an uptake in the R.E. tissues would seem to indicate that it was not. Our first batch was obtained from Philips and I believe it was supplied as a citrate. There has been a lot of work done by van der Werf on the use of ^{206}Bi in a colloidal form as well as in soluble form.

W.H. BEIERWALTES: I would like to reinforce Glenn's remarks on the problem of whether the compound that you are going to inject into an animal or, for example, into humans is really in solution. Formulation is indeed a big problem. After we had actually imaged the dog adrenals with ^{125}I-19-iodocholesterol, we ran into difficulties in preparing a useful ^{131}I-19-iodocholesterol. When we mixed iodocholesterol with various solutions, we sometimes got precipitates and at other times found particles that disappeared when the solution was heated. The observation of a high uptake of a radiolabelled compound in the lungs or in the liver tells you that you don't have a solution but a suspension, whether you have seen it or not. As to iodocholesterol, it's absolutely critical that it must be in 0.2% solution of Tween-80 in ethanol and an exact percentage of saline, otherwise it just doesn't work.

R.L. HAYES: Let me add one short comment on the question of iron. It was very tempting, even if it was not feasible clinically, to speculate that, since there was binding of gallium to transferrin, one could simply block this binding by giving sufficient ionic iron, for instance by giving ferric ammonium sulphate, or something like that. Our original animal experiments were along that line and they indicated that the tumour-to-non-tumour ratio was increased but not dramatically so. That is what makes the behaviour of scandium even more remarkable. On the basis of some other experiments by Dr. Hartman, we conjectured that scandium was binding to more than just transferrin. The reason for this is that in in-vitro experiments, where the iron-binding capacity of serum has been determined, the addition of enough iron to saturate the iron-binding capacity of the serum produced a drop in the normal binding of gallium from about 95% to about 20%.

A doubling of the iron added to the serum did not produce any further drop in the binding of gallium to the serum indicating that although transferrin might be the major species that binds gallium, it was still not responsible for all the binding. The idea of being able to enhance the distribution of a particular material came at a time when we wanted to block the binding of gallium to plasma proteins in order to promote the deposition of gallium in bone. It seemed very logical to use iron experimentally, but not clinically. We already knew at that time that gallium alone would produce a block and it is very obvious why it would. But the amount of gallium required to obtain satisfactory blocking was quite high and objectional from a toxicity standpoint. On the other hand, the amount of scandium required was very low.

D. COMAR: I think we can make a general statement. We still do not know why all these elements are taken up by tumours, and it was a kind of chance when we started looking into the uptake of gallium, indium and mercury. We have a number of examples in our laboratory where we have tried for rational reasons to use labelled compounds as tumour-localizing agents, but at least in some of the cases it didn't work the way we thought. Still, we depend quite a bit on fortune.

RADIOIODINE-LABELLED COMPOUNDS PREVIOUSLY OR CURRENTLY USED FOR TUMOUR LOCALIZATION

W.H. BEIERWALTES
Section of Nuclear Medicine,
University of Michigan,
Ann Arbor, Michigan,
United States of America

Abstract

RADIOIODINE-LABELLED COMPOUNDS PREVIOUSLY OR CURRENTLY USED FOR TUMOUR LOCALIZATION.
^{131}I-labelled human serum albumin, though not used for tumour localization today, is an excellent "standard" with which to compare uptake of "tumour-specific" radiolabelled compounds. ^{131}I-labelled fibrinogen and antibodies to fibrinogen have a non-specific uptake in tumours. Nungester, Beierwaltes and Knorpp are credited by Mahaley as first treating a human for cancer with ^{131}I-labelled antibody globulins (malignant melanoma). Although many theoretical problems remain in obtaining diagnostic localization of ^{131}I-IgG, Quinones, Mizejewski and Beierwaltes demonstrated the uptake of ^{131}I-labelled immune antibodies in Syrian hamster cheek pouch with chorionic gonadotropic hormone as the specific tumour-associated antigen. This model was then used successfully by Goldenberg and Hoffer for demonstrating colon carcinoma by using antibodies to carcinoembryonic antigen. A ^{131}I-labelled chloroquine analogue, synthesized by Counsell, has been demonstrated by Beierwaltes et al. to concentrate diagnostically and therapeutically in the malignant melanotic melanoma. ^{131}I-19-iodocholesterol, synthesized by Counsell, has been demonstrated by Beierwaltes et al. to concentrate diagnostically in the human adrenal cortex. It has many unique diagnostic capabilities not available with other routine diagnostic methods available today.

^{131}I-LABELLED HUMAN SERUM ALBUMIN

Apparently, brain tumour scanning was the first use of radioiodine-labelled human serum albumin for tumour scanning [1]. There are at least five reasons why this substance might show an increased concentration in tumours [2]: increased vascularity, abnormal vascular permeability, enlarged extracellular space, reactive oedema and cellular metabolism. To date, no substance has been found which actively concentrates in brain tumours or achieves an ultimate tumour concentration greater than the initial blood level. Therefore, the seemingly high uptake in brain tumours is a relative phenomenon, attributable to the almost uniquely low radioactivity background of normal brain.

Radioiodinated human serum albumin has largely been replaced by ^{99}Tc^m as a brain-tumour scanning agent because of the necessity to wait 24 hours for scanning to allow a decrease in blood background, unnecessarily high radiation dose, etc.

^{131}I human serum albumin uptake in tumours in general, however, is an excellent standard against which to measure "specific" uptake of the radioiodine-labelled compounds. For example, in our work with malignant melanotic melanomas in the hamster with ^{125}I-chloroquine analogue [3], the tumour-to-blood and tumour-to-liver ratios were as high as 4 : 1 and allowed imaging of the melanoma with rectilinear scintillation scanning.

Intra-arterially injected macroaggregates of albumin may be thought of as adjuncts to tumour angiography rather than as tumour scanning agents.

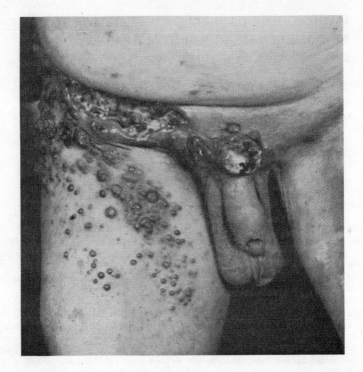

FIG.1a. *Right groin of 63-year-old male on 11 Dec. 1951 when he was given 27 mCi of ^{131}I-labelled rabbit globulin "immune" to a crude extract of his malignant melanoma.*

^{131}I-FIBRINOGEN

Radioiodinated fibrin and fibrinogen accumulate in most solid tumours [4]. A possible mechanism is that all solid tumours undergo necrosis and haemorrhage with deposition of fibrin-containing blood clots. The uptake of radioiodinated fibrinogen and fibrin is therefore non-specific and should be greatest late in the development of tumours.

More recent efforts centered upon the development of antibody preparations with highest specific activity to fibrin and fibrinogen. Spar et al. [5] produced ^{131}I concentrations in transplantable rat tumours as high as 15% dose/g using radioiodinated gamma globulin from immunized rabbits as compared to a 2% dose/g using ^{131}labelled non-immunized rabbit gamma globulin. McCardle et al. [6] gave 200 – 800 µCi of ^{131}I-labelled and purified rabbit antibodies to human fibrogen to 50 patients, without reactions. They achieved tumour-to-blood pool radioactivity for 10 days after injection and scan localization in 58% of patients. They gave treatment doses of ^{131}I using this technique with 150 mCi of antibody delivering a calculated 2 000 rads.

^{131}I-GLOBULINS

Most of the interest in tumour localization with ^{131}I-labelled globulins (and globulin subunits) has centered around their sources of antibodies to possible tumour-specific antigens.

Nungester, Beierwaltes and Knorpp are credited by Mahaley [28] for the first attempt to treat a human cancer with ^{131}I-labelled antibodies. A 63-year-old male had a malignant melanoma recurrent

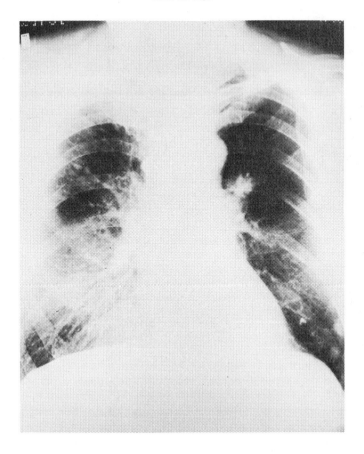

FIG.1b. Chest X-ray on 11 Dec. 1951 showed approximately seven lung metastases.

in the right inguinal area repeatedly after previous surgery and X-irradiation. On 10 Oct. 1951 tumour tissue was removed, an extract made, and rabbits were immunized with the extract with complete Freund's adjuvant. On 11 Dec. 1951 the patient was given 27 mCi of ^{131}I tagged to $\cong 90$ mg of rabbit immune globulin. At this time the patient had seven visible lung metastases and an estimated 85 skin metastases, with two fungating inguinal ulcers (Figs 1a and 1b). By 3 Jan. 1952 a 40% reduction in the groin lesions was visible and some decrease in lung lesions had occurred. He was given a repeat treatment on 4 Jan. 1952. By 20 May 1952, no tumour remained (by biopsy of dermal dark spots) (Fig. 2a) and the chest X-ray was normal (Fig. 2b). He died 9 years later of a coronary occlusion. An autopsy disclosed no residual neoplasm. Unfortunately, subsequently, 13 similar patients were similarly treated without beneficial results.

In theory, a pure tumour-specific antigen could be combined with a suitable adjuvant and used to produce a tumour-specific antibody without the simultaneous production of antibody to other normal human tissues. The principal difficulties are:

1. It is extremely difficult to purify an active tumour-specific antigen.
2. Even when the IgG fraction of the gamma globulin is isolated, only 1% or less of that fraction is specific for that tumour. Obviously, the initial antibody titre in the gamma globulin

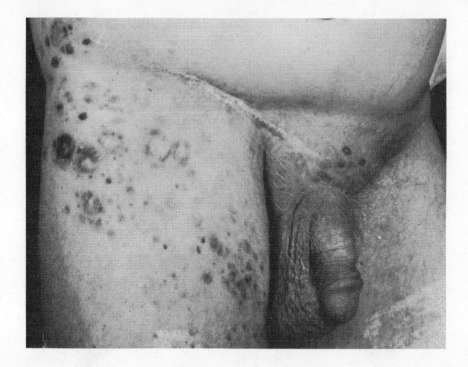

FIG.2a. Appearance of right groin in same patient on 20 May 1952 when rebiopsy of black lesions showed no residual melanoma.

fraction would have to be very high to result in a diagnostically helpful antibody titre after separation of the IgG fraction, at least 99% of which is non-specific for the tumour.

3. Tumour antigen usually leaks into the circulation and may combine with endogenous antibody producing a "blocking" antibody, or may combine with exogenously administered antibody before it reaches the tumour and prevent its diagnostic uptake.

In spite of these limitations, diagnostic specific tumour localization has been achieved. Quinones, Mizejewski and Beierwaltes [7] immunized rabbits to human chorionic gonadotropic hormone (hCGH) as a tumour-specific hormone in the male with choriocarcinoma. Radioiodinated IgG from the immunized rabbit globulin concentrated both significantly and specifically more than radioiodinated IgG from non-immunized rabbits in human choriocarcinoma in vitro and in vivo in the Syrian hamster cheek pouch. This concentration was sufficient to allow diagnostic scanning of the choriocarcinoma.

The success of this approach has been confirmed and extended by Hoffer et al. [8] using ^{131}I-carcinoembryonic antigen (CEA). Hoffer [8] demonstrated that radioiodine-labelled IgG against human carcinoembryonic antigen from immunized goats was superior to ^{111}In-chloride, ^{67}Ga-citrate and ^{111}In-bleomycin for localization of colonic carcinoma in the Syrian hamster cheek pouch. He believes that it will be useful in primary lesions and hepatic metastases from colonic tumours. He also demonstrated diagnostic localization in one human with carcinoma of the tonsil and cervical lymph node metastases [9].

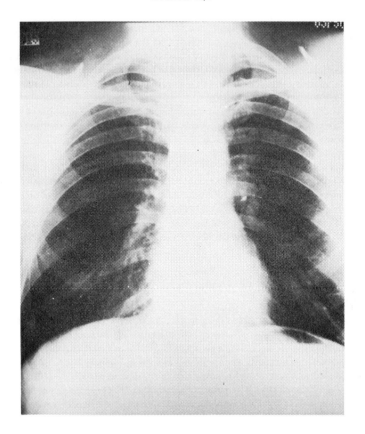

FIG.2b. *Chest X-ray on 20 May 1952 is normal.*

^{131}I-SYNKOL

The first radioiodine-labelled antitumour agent known to localize differentially in tumour tissue was ^{131}I-labelled Synkol 2 : 3-dimethyl-5 : 6-di-^{131}I-iodo-1 : 4 benzohydroquinone diphosphate, a radiosensitizing antitumour agent [10]. This agent has met with limited success due, in part, to its relatively low per cent uptake in dose/g and a tumour-to-liver ratio of usually <1 [11].

^{131}I-CHLOROQUINE ANALOGUES

In 1966, Drs Counsell and Morales and I noted that chloroquine, used to treat malaria, occasionally caused a retinopathy. Several reports indicated that chloroquine was concentrated in melanin in the choroid of the eye and was released slowly.

We therefore evaluated the uptake of ^{14}C-chloroquine in mice with melanomas and found that the uptake and retention of chloroquine in melanomas was similar to that of choroid [3].

Dr. Counsell [12] synthesized a radioiodinated analogue of chloroquine 4-(3-dimethylaminopropylamino)-7-iodoquinoline, hereafter called NM 113. We demonstrated that this ^{125}I-labelled agent concentrated diagnostically in dermal [11, 13-15] and ocular [16-17] melanotic melanomas.

By using therapeutic doses of ^{131}I-labelled NM 113 in dogs, ~ 3 200 rads of irradiation was demonstrated to be delivered to melanoma metastases. One dog experienced a total disappearance of both the recurrent primary and the cervical node metastases between the 17th and the 30th day, with no detectable change in retinal function [18]. One human given a 100-mCi dose experienced a temporary partial regression of recurrent melanoma metastases during the same period of time after the dose [19].

More recently, we have evaluated a new radiolabelled chloroquine analogue, 4 amino (benzo-4-methylpiperazine)-7-trifluoromethyl quinoline-^{3}H, (and a radioiodinated analogue of this compound) which concentrates in the malignant melanoma of mice 4 X > ^{14}C-chloroquine and 3 X > ^{125}I-NM 113 which had a 3 X greater specific activity [20].

The principal limitations of the diagnostic use of ^{125}I- and ^{131}I-NM 113 are that the high concentration in normal lung and the excretion from liver in bile to intestines limits its use in localizing dermal metastases in the thoracic and abdominal areas. The principal limitation of its use in the diagnosis of ocular melanomas is the high background radioactivity in normal choroid. Currently, we hope to solve the latter problem with the simultaneous use of a directional ultrasound probe to localize the tumour and a new focused intraocular radiation detector to prove that the tumour is a melanotic melanoma [21].

RADIOIODINATED o, p' DDD ANALOGUE AND CHOLESTEROL AS ADRENAL TUMOUR-IMAGING AGENTS

Our first published efforts [22] at imaging the adrenal were with a radioiodinated DDD analogue, o, p' DDD [1, 1-(p-iodophenyl)-2, 2-dichloroethane]. Although we did image the dog adrenal in vivo, the image was faint. The principal limitation of this compound (an adrenal cortex enzyme inhibitor) is that although it concentrated in the dog adrenal cortex 2 X that in liver at 4 hours, the concentration in fat progressively rose to 4 X that in adrenal by 24 hours.

Appelgren [23] demonstrated by autoradiography that ^{14}C-cholesterol concentrated strikingly and best in the adrenal cortex of mice. Nagai et al. [24] imaged the human adrenal faintly with ^{131}I-stigmasterol. Dr Varma and I [25] demonstrated that ^{14}C-cholesterol concentrated in the adrenal cortex of the dog in 0.6% dose/g. Counsell et al. [26] synthesized ^{125}I-19-iodocholesterol. In a series of subsequent publications [27] we have demonstrated that ^{131}I-19-iodocholesterol is most helpful in: (1) Differentiating Cushing's syndrome caused by ACTH excess from the hyperfunctioning adrenal cortical adenoma. (2) In lateralizing aldosteronomas. (3) Is *uniquely* useful in detecting post-adrenalectomy remnants in patients with persistent cortisol excess, in diagnosing androgen-secreting and cortisol-excess syndromes before they can be diagnosed by conventional methods, in diagnosing androgen-secreting adrenal cortical carcinomas, in patients who need contrast venography but who are allergic to contrast media or where the adrenal vein cannot be catheterized, or in detecting adrenal nodules or adenomas missed by venography. The procedure possesses the additional advantages over venography of being non-traumatic, non-invasive and not requiring hospitalization.

The principal limitations are that the radiation dose to the gonads is similar to that of an intravenous pyelogram and it takes days to complete the study.

REFERENCES

[1] MOORE, G.E., Use of radioactive diiodofluorescein in the diagnosis and localization of brain tumors, Science **107** (1948) 509.

[2] BAKAY, L., Basic aspects of brain tumor localization by radioactive substances: A review of current concepts, J. Neurosurg. **27** (1967) 239.

[3] BEIERWALTES, W.H., et al., Scintillation scanning of malignant melanomas with radioiodinated quinoline derivatives, J. Lab. Clin. Med. **72** (1968) 485.

[4] DAY, E.D., et al., Localization in vivo of radioiodinated rat fibrinogen in the Murphy rat lymphosarcoma and in other transplantable rat tumors, J. Natl. Cancer Inst. 22 (1959) 413.
[5] SPAR, I.L., et al., Preparation of purified ^{131}I-labelled antisera to human fibrinogen — preliminary studies in human patients, Acta Un. Int. Cancer 19 (1963) 197.
[6] McCARDLE, R.J., et al., Studies with iodine-^{131}I-labeled antibody to human fibrinogen for diagnosis and therapy of tumors, J. Nucl. Med. 7 (1966) 837.
[7] QUINONES, J.D., et al., Choriocarcinoma scanning using radiolabeled antibody to chorionic gonadotrophin, J. Nucl. Med. 12 (1971) 69.
[8] HOFFER, P.B., et al., Use of ^{131}I-CEA antibody as a tumor scanning agent, J. Nucl. Med. 15 (1974) 323.
[9] HOFFER, P.B., et al., Tumor scanning agents, Semin. Nucl. Med. 4 (1974) 305.
[10] MARRIAN, D.H., MAXWELL, D.R., Tracer studies of potential radiosensitizing agents: Tetrasodium 2-methyl-3-^{82}Br-bromo-1 : 4-naphthhydroquinone diphosphate and tetrasodium 2 : 3-dimethyl-5 : 5-di^{131}I-iodo-1 : 4-benzohydroquinone diphosphate, Br. J. Cancer 10 (1956) 739.
[11] GANATRA, D., et al., "Radioiodinated synkol as a tumour-localizing agent", Medical Radioisotope Scintigraphy (Proc. Symp. Salzburg, 1968) 2, IAEA, Vienna (1968) 25.
[12] COUNSELL, R.E., et al., Tumor localizing agents. III. Radioiodinated quinoline derivatives, J. Pharm. Sci. 56 (1967) 1042.
[13] BEIERWALTES, W.H., et al., Scintillation scanning of malignant melanomas with radioiodinated quinoline derivatives. Preliminary note, J. Nucl. Med. 9 (1968) 489.
[14] BEIERWALTES, W.H., et al., Visualizing human malignant melanomas and metastases with radioiodinated quinoline. Use of chloroquine analog tagged with iodine-125, J. Am. Med. Assoc. 206 (1968) 97.
[15] BOYD, C.M., et al., Diagnostic efficacy of radioiodinated chloroquine analog in patients with malignant melanomas, J. Nucl. Med. 11 (1970) 479.
[16] BOYD, C.M., et al., ^{125}I-labeled chloroquine analog in the diagnosis of ocular melanomas, J. Nucl. Med. 12 (1971) 601.
[17] KNOLL, G.F., A gamma-ray probe for the detection of ocular melanomas, IEEE Trans. Nucl. Sci. NS 19 (1972) 76.
[18] LIEBERMAN, L.M., et al., Treatment doses of ^{131}I-labeled chloroquine analogue in normal and malignant melanoma dogs, J. Nucl. Med. 12 (1971) 153.
[19] BEIERWALTES, W.H., "Labeled chloroquine analog in diagnosis of ocular and dermal melanomas", Proc. Hahnemann Symposium on Diagnosis of Tumors with Radionuclides, to be published by Wiley & Sons Publ. Co.
[20] BEIERWALTES, W.H., et al., A new radiolabeled quinoline analog in mice with malignant melanomas, Presented at 47th Annual Meeting of the Central Society for Clinical Research (Abstract in Clinical Research 22 (1974) 640A).
[21] ROGERS, W.L., et al., "An ultrasound-guided gamma-ray probe for detection of ocular melanomas," 1974 Ultrasonic Symposium Proceedings, IEEE Catalogue #74CHO-896-ISU (IEEE Ultrasonics Symposium Milwaukee, Nov. 11–15, 1974).
[22] Di GIULIO, W., BEIERWALTES, W.H., Tissue localization studies of a DDD analog, J. Nucl. Med. 9 (1968) 634.
[23] APPELGREN, L.E., Sites of steroid hormone formation. Autoradiographic studies using labeled precursors, Acta Physiol. Scan. 71 301 (1967) 1.
[24] NAGAI, T., et al., An approach to developing adrenal gland scanning, J. Nucl. Med. 9 (1968) 576.
[25] BEIERWALTES, W.H., et al., Per cent uptake of labeled cholesterol in adrenal cortex, J. Nucl. Med. 10 (1969) 387.
[26] COUNSELL, R.E., et al., Tumor localizing agents IX radioiodinated cholesterol, Steroids 16 (1970) 317.
[27] SEABOLD, J.E., BEIERWALTES, W.H., "Adrenal imaging", in Nuclear Medicine in Clinical Practice (SCHNEIDER, D.B., TREVES, S., Eds), ASP Biological and Medical Press B.V., Amsterdam, to be published in 1976.
[28] MAHALEY, M.S.J., Immunological considerations and the malignant glioma problem, Neurosurgery 15 (1968) 178.

DISCUSSION

H.J. GLENN: A brief remark on the availability of these labelled compounds seems very appropriate at this point. When compounds are available they become widely used, but when they are not available they stay very much in the research field and, important or not, they simply don't get the wide use that some of them deserve. This is true for the antigen/antibody preparations, and the same has been the case for such a simple compound like labelled fibrinogen, a parenteral on which the commercial companies have been working for several years. Because of the hetatitis scare and the requirements laid down by regulatory agencies, the availability of this substance has been greatly delayed.

R.M. KNISELEY: We have been talking mostly about ^{131}I. It now seems as though Los Alamos is able to produce curie amounts of ^{123}I, which have some real advantages over ^{131}I in terms of the radiation dose. I would like to have some comments on this point. As to the iodine-labelled cholesterol studies, you still might not find ^{123}I a good choice, even though you could perhaps use ten times the dose with less exposure to the patients. Is that correct?

W.H. BEIERWALTES: Initially, we didn't use ^{123}I because of the short shelf-life and the long time required for imaging the adrenals with iodocholesterol. However, the new 6-beta-19-norcholesterol images the adrenal in one day.

R.M. KNISELEY: Who is familiar with the potential, availability and cost of ^{123}I?

R.L. HAYES: Los Alamos will be able to produce hectocuries which would be enough for the whole world.

H.J. GLENN: At meetings in the last five or more years we have heard from many different groups that they now have a cyclotron process that would give us large quantities of ^{123}I. Essentially none of these processes have paid off from a commercial point of view. I wonder how the short half-life of ^{123}I would influence the effectiveness of your iodocholesterol.

W.H. BEIERWALTES: One of our problems was the length of synthesis and then the two weeks that we want to follow the patient. It now looks as if the most critical period for studying the patient is during the first three to five days, and particularly with suppression scanning. This now brings back into the realm of practicality the possibility of using ^{123}I. As to the availability and price of ^{123}I, Mediphysics is trying for a new drug application (NDA). Even if ^{123}I is considerably more expensive than ^{131}I, the importance of the decreased radiation dose has strong appeal.

V.R. McCREADY: We have regular supplies of ^{123}I from Hammersmith each week and we routinely use ^{123}I-Hippuran for children. It is possible to produce ^{123}I on a larger scale in England, presumably for the whole of Europe. With increased usage the price would go down to very cheap levels as it did with ^{18}F, unless there is some basic physical reason why it can't. I have been told that for example ^{67}Ga would always be expensive.

D. COMAR: In Hammersmith ^{123}I is prepared by bombarding antimony with alpha particles. The ^{123}I isn't pure, it contains a few per cent of ^{124}I and also a little ^{125}I. At the time of delivery the radioisotope may be quite pure, and for Hippuran examinations it is of no importance. But for labelling cholesterol you have to wait many days before obtaining the results and the high amounts of ^{123}I would decrease quickly and the ^{124}I would at this time maybe account for ten times higher activity than your ^{123}I. So this method of preparing of ^{123}I is not very good for such studies. Another way of making ^{123}I is by bombarding tellurium with protons, I think. But the problems are the same as when you use antimony with alpha particles; also in this case you will have a few per cent of ^{124}I or ^{125}I. The best way of producing pure ^{123}I is by bombarding ^{127}I with high-energy particles (protons or deuterons) which we can't obtain with small cyclotrons. We must be very careful when working with short-lived radionuclides because there often will be some contamination, perhaps very low, but we must take it into account. In addition, we have the problems about delivery. I don't know the exact details for ^{123}I, but for other isotopes made in the USA it sometimes takes a week before they arrive in Paris, as for instance with ^{67}Cu (a $2\frac{1}{2}$-day isotope), which we buy for the moment from Oak Ridge. Even if an isotope is available, difficulties in the transportation may hamper the general use of this nuclide.

W.H. BEIERWALTES: I remember one of the reasons why we decided against the use of ^{123}I was that after 24 hours the radiation dose from ^{124}I becomes equal to the dose that you would have from ^{131}I. The people in Miami say that they don't use ^{123}I after 24 hours, they throw it away. They either get the ^{123}I daily or they don't use it at all.

E.H. BELCHER: Could I ask Dr. Beierwaltes if he could briefly indicate the metabolic fate and the biological half-life of the radioiodine labels in the labelled cholesterol and the other labelled compounds that he was talking about?

W.H. BEIERWALTES: Sitting on the shelf there is a 7% deiodination of radioiodine when it is in the refrigerator. If it is kept at room temperature, the deiodination exceeds that, I believe it is 17% deiodination at two weeks. If the ^{131}I-19-cholesterol doesn't work we are looking for one of two explanations: The first explanation is that you may find 90% of the activity at the injection site even if you think that the injection was perfect. Another explanation may be that the radioiodinated cholesterol was deposited in the receiving place at 25°C for a week. In this case I suggest you put your camera over the patient's thyroid to see where the radioiodine has gone. Those are the most common causes for lack of adrenal imaging. As to the excretion I believe that something like half of the radioiodine is excreted in the urine and half in the stools. The peak excretion in the urine occurred at 24 hours and that in the stools at 28 hours. The biological half-life in the human was about 2.15 days and the effective half-life somewhat more than 2 days.

K. HISADA: Dr. Kojima, Kyushu University, and Dr. Ogawa and their associates in the Daichi Radioisotope Labs. Ltd. in Japan have recently synthesized a new compound by changing the position of the iodine label in the cholesterol. They call the new compound adosterol. In Japan we are now switching from iodocholesterol to adosterol in clinical institutions. By means of this compound, 6β-iodomethyl-19-nor-cholest-5(10)-en-3β-ol-^{131}I (NCL-6-^{131}I), we can reduce the radiation dose to the patients.

W.H. BEIERWALTES: I am glad that you mention this compound. In our early work in producing ^{131}I-19-iodocholesterol we were very bothered by an impurity that appeared in the iodocholesterol. We had already seen that impurity earlier in our experiments with dogs and had been successfully imaging the dog adrenals with the impurity. So when we once more had the impure compound, we gave it to the patients that day and we got the best images we ever had. Since then it turned out that this impurity is 6β-19-nor-iodocholesterol, which actually changes some double bonds and is better than ^{131}I-19-iodocholesterol.

T. MUNKNER I wonder whether you have found impurities in the ^{123}I-iodide preparations. We had troubles in using the commercial preparation for labelling purposes.

D. COMAR: Do you mean stable iodide mixed with your ^{123}I-iodide or tracers of the target material, which is either antimony or tellurium?

T. MUNKNER: Our target material was antimony. We looked for traces of the target material by proton activation of the material, but we only found traces of different metals. At least it made our labelling or exchange of iodine in Hippuran impossible.

D. COMAR: If you produce ^{123}I by high-energy proton bombardment of ^{127}I, you obtain ^{123}Xe from which you obtain your final ^{123}I. In this case you are supposed to have a very pure product, not containing any carrier except a little ^{125}I ($\cong 0.1\%$ EOB).

When I think about antibodies labelled with short-lived isotopes, I am troubled by the time necessary for the uptake of antibodies in tumours. Is this relatively long time due to impurities in the antibodies which make it necessary to wait for the elimination of those labelled compounds which are not taken up by the tumour, or is it due to a metabolic process responsible for the uptake of antibodies in the tumour?

W.H. BEIERWALTES: This question I can't answer exactly, but it is our impression that the higher the specific activity of the radiolabelled compound, generally speaking, the higher the concentration in the tissues of interest will be, and also the faster the uptake. I would like to point to a parallel situation in radioimmunoassay procedures where it may take up to five days of incubation or more before you get exactly the type of curve you want. That means that under in-vitro conditions you have such antibody/antigen combinations that require an extended time before you have a reasonable completion of the reaction.

V.R. McCREADY: You didn't comment on the possibilities of differential diagnosis using radiolabelled albumin. I am referring to the work done by Mme Planiol, who compared the uptake of iodine-labelled albumin in the two brain hemispheres giving the results as a ratio. Doing so, she found different tumours having different ratios of accumulation of the nuclide, and her diagnostic accuracy was quite good. Presumably the different accumulation rates reflect the variation in tumour permeability and vascularity.

W.H. BEIERWALTES: Listening to the comments at the meeting on the uptake of radio-iodine-labelled human serum albumin in tumours, I think it may very well be a fruitful investigation to study the incorporation of labelled albumin in different tumours. I have also been very impressed with the differential concentration of radioiodinated human serum albumin in malignant melanomas as compared to liver ratios. In fact I would defy anyone who thinks he has a good compound to routinely check against the uptake of radioiodinated human serum albumin in that specific tumour.

V.R. McCREADY: My colleagues working in tumour immunology tell me that the antigenic properties of tumours are extremely variable. However, Dr. Beierwaltes, you have a great experience in this area.

If you make a melanoma antigen as you did for one human, do you expect it to have the same antigenic properties as a melanoma in another human? And do you think that the antigenic properties change from tumour to tumour within the human species?

W.H. BEIERWALTES: I expect both. In our work with irreversible enzyme inhibitors we have found that the irreversible enzyme inhibitor may be extremely specific, and in some cases not only the right enzyme, but also the right isoenzyme is necessary to show the type of concentration that we would like to apply for diagnostic uses. Similarly, there is some evidence that the antibody may be person specific. It doesn't bother us because if the time interval is not too long, we would be very glad to take advantage of the specificity for that person and prepare a unique antibody for each individual. But we know that you can also have irreversible enzyme inhibitors or antibodies that are broader in their application and will apply to the tumour regardless of the individual. Recently, there has been a publication in the New England Journal of Medicine where the workers had prepared an antigen, I guess from the cell membrane fraction of an ocular melanoma, and they were able to produce a characteristic delayed skin reaction in everyone with an ocular melanoma, but not in patients with lesions simulating ocular melanomas. In the cases which were positive, the ocular melanomas were down to 2 – 3 mm in diameter.

V.R. McCREADY: Would you like to give a general statement about this immunological approach: is it a good one to follow? Are you hopeful that it will pay off?

W.H. BEIERWALTES: We are following this approach.

IAEA-MG-50/12

TUMOUR LOCALIZATION USING RADIOMERCURY-LABELLED COMPOUNDS

C. RAYNAUD, D. COMAR
CEA, Département de biologie,
Service hospitalier Frédéric Joliot,
Orsay, France

Abstract

TUMOUR LOCALIZATION USING RADIOMERCURY-LABELLED COMPOUNDS.
Radioactive mercury, either in the form of chlormerodrin or $HgCl_2$, is used to detect tumours. ^{197}Hg-chlormerodrin is used particularly in studying brain tumours and $^{197}HgCl_2$ is used essentially in studying lung tumours. The physical half-life of 65 hours for ^{197}Hg allows early and late scans and its use in nuclear medical services far from production centres. These advantages partially compensate for the dosimetric disadvantages. The use of ^{203}Hg should be restricted because of the high radiation dose absorbed by the kidneys.

Radioactive mercury was proposed for tumour detection first by Blau and Bender [1] in 1962 in the form of chlormerodrin and then by Wolf and Fischer [2] in 1964 in the form of mercuric chloride. These two compounds labelled with ^{197}Hg have been widely used in the last decade to detect a large variety of tumours.

CHLORMERODRIN

Chlormerodrin is a mercurial diuretic which was first used labelled with ^{203}Hg and then, as suggested by Sodee [3], labelled with ^{197}Hg. When it is injected intravenously, the plasma activity decreases quickly, since 70 to 75% of the injected dose is eliminated in urine during the first 24 hours. Renal uptake increases to reach about 8% of the injected dose on the second day and remains at this level for several days. Liver uptake also increases but remains at a much lower level. For the other parenchymas the Hg concentration is low and decreases with time.

The tumour concentration of the Hg injected in the form of chlormerodrin increases in the first few hours and reaches a plateau during the 1st or 2nd day. The Hg seems to be particularly linked with the cytoplasmic proteins of the tumour cells [4, 5]. The divergence of the curve for tumour activity, which is shown as a plateau, from the curve for tissue activity, which decreases, indicates that the best images will be obtained late after injection.

^{203}Hg- or ^{197}Hg-chlormerodrin has been used to diagnose tumours of the face and neck [3, 7], breast [3], prostate [3], bones [3], eyes [8], stomach and intestine [3], lung [3] and, above all, the brain [1, 2].

Twelve years after the publication of the paper by Blau and Bender [1], ^{197}Hg-chlormerodrin and, to a lesser extent, ^{203}Hg-chlormerodrin are still being used to detect brain tumours. The possibility to obtain early and late images with this compound gives it a considerable advantage over substances labelled with $^{99}Tc^m$. However, because of its accumulation in the kidneys, the radiation dose absorbed by the kidneys during examination with 750 µCi of ^{203}Hg-chlormerodrin is very high, about 165 rads [9]; this means that the use of this product should be restricted to an indispensable minimum. With ^{197}Hg-chlormerodrin, the dose absorbed by the kidneys for 750 µCi is 13 rads [9].

^{197}Hg-chlormerodrin is also used to detect kidney tumours but its use is quite different. The renal cortex normally accumulates about 8% of the dose injected. When a renal tumour develops and invades the cortex, such physiological uptake decreases or disappears and the tumour is indicated by a defect on the scintigram. The tumour concentration of Hg is negligible compared with the high uptake in the normal tissue. This technique does not differentiate between benign and malignant tumours.

MERCURIC CHLORIDE

The metabolism of $HgCl_2$ is quite different from that of chlormerodrin. After intravenous injection of $^{197}HgCl_2$ the plasma activity decreases slowly, the daily urine elimination is very small (about 0.5% of the dose) and the faecal excretion is high and reaches 4 to 5% of the dose during the first days. Renal uptake is high, representing about 20% of the injected dose on each side, and the biological half-life of the Hg in the kidney is approximately 100 to 120 days. Liver uptake is lower and does not seem to last as long. Tissue uptake for parenchymas, other than the kidneys, the liver and the spleen, is low.

Tumour accumulation of Hg injected in the form of $^{197}HgCl_2$ has been observed in transplanted Ehrlich ascites cell carcinoma of mice [10] in which the tumour-to-supernatant ratio is, on average, about 7.2 at the 96th hour. For $^{111}InCl_3$ and ^{67}Ga, this ratio is 1.85 and 0.70 respectively [10] and is much less favourable. In man, the tumour concentration of Hg has been found to be approximately 0.25 and 1% of the injected dose in two cases of superficial malignant tumours [11, 12]. Such tumour concentrations can only be detected in cases in which they are clearly higher than the activity of the parenchyma on which the tumour has developed. Consequently, $^{197}HgCl_2$ has been used to detect tumours of the face and neck [2, 13], breast [11, 14], thyroid [15], brain [2] and, above all, the lungs [2, 11, 16—22].

Now, a decade after Wolf's and Fischer's initial publication [2], the detection of lung tumours by means of $^{197}HgCl_2$ has become a routine technique [2, 11, 16—22].

However, the tumour uptake of $^{197}HgCl_2$ is no more specific for the cancer than is that of chlormerodrin or other radioactive compounds used for the same purposes, such as ^{67}Ga- or ^{57}Co-bleomycin [23]. A good scintigraphic technique allows the detection of all cancers if they have a diameter of more than 2 cm, a solid consistency and have not been treated by chemotherapy or radiotherapy: there is no detectable uptake in benign tumours but unfortunately $^{197}HgCl_2$ uptake in active inflammatory lesions may be as high as in cancer [23—25]. Despite this lack of specificity, $^{197}HgCl_2$ remains useful to the clinician in differentiating benign tumours from malignant tumours to determine the mediastinal extension of cancer and to detect as early as possible a recurrence in a patient already operated on, as well as in cases of nodular shadows [17].

As with chlormerodrin, $^{197}HgCl_2$ can be used to detect kidney tumours. A defect on the scan may indicate the presence of a tumour, but a cyst and a malignant tumour cannot be differentiated.

High renal uptake of Hg injected in the form of $^{197}HgCl_2$ excludes the use of ^{203}Hg. With ^{197}Hg, the radiation dose absorbed by the kidneys is about 30 rads for an injection of 1000 μCi [25].

OTHER MERCURIC PRODUCTS

Tumours of the spleen can be detected by using a mercuric compound, ^{197}Hg-MHP (1-mercuri-2-hydroxypropane) [27].

Here, as with renal tumours, a tumour is indicated by a defect on the scan. Healthy spleen tissue sequestrates red cells damaged by ^{197}Hg-MHP, whereas tumourous tissue loses this property. The radiation dose absorbed by the kidneys is 7 rads for 300 μCi of ^{197}Hg-MHP [9].

Another mercuric compound, ^{197}Hg-acetate, has the same properties as ^{197}HgCl$_2$ and can be used with the same indications [23]. It is possible that this compound can be used clinically in smaller doses and, consequently, the radiation dose absorbed by the kidneys be reduced.

DISCUSSION

At a time when, under the influence of ecologists, considerable effort is being made to inform the public about the dangers of Hg contamination, it is important to examine in detail the possible harmful effects of radioactive Hg preparations used.

With ^{197}Hg-chlormerodrin[1] and a usual dose of 750 μCi, the radiation dose absorbed by the kidneys is moderate, reaching 13 rads [9]. However, the stable Hg load is not negligible and represents 1000 to 2000 μg of which only 300 to 600 μg remain in the body after 24 hours. For comparison, we should remember that our daily food intake introduces into the body 5 to 25 μg of Hg [28].

With ^{197}HgCl$_2$[1] and a standard dose of 1 mCi, the radiation dose absorbed by the kidneys is approximately 30 rads [26] and cannot be neglected. However, the amount of stable Hg, ranging between 5 and 20 μg, is very small and barely represents the amount introduced into the body by daily food intake.

These disadvantages explain the preference for ^{67}Ga and compounds labelled with ^{99}Tc^m. However, the physical half-life of 65 hours and the low cost of ^{197}Hg which, for example, in the case of ^{197}HgCl$_2$ is 15 times less than that of ^{67}Ga, remain important advantages for some users.

These advantages and disadvantages are weighted differently depending on the nuclear medicine centre involved. ^{197}Hg is especially interesting in less favoured countries and in laboratories far from centres of radioisotope production. The total amount of ^{197}Hg sold at the present time by CEA, IRE and SORIN for tumour detection may give an idea of the consumption of ^{197}Hg for this purpose: it is about 2.5 Ci per month (30% being ^{197}Hg-chlormerodrin and 70% ^{197}HgCl$_2$).

CONCLUSION

^{197}Hg is used in detecting tumours either in the form of chlormerodrin or in the form of HgCl$_2$. The first of these two compounds is used particularly to detect brain tumours and the second to detect lung tumours. Despite some disadvantages, ^{197}Hg compounds are widely used.

REFERENCES

[1] BLAU, M., BENDER, M., Radiomercury (^{203}Hg) labelled Neohydrin: a new agent for brain tumor localization, J. Nucl. Med. 3 (1962) 83.
[2] WOLF, R., FISCHER, J., "Szintigraphische Untersuchungen mit ^{197}HgCl$_2$", 45th Dtsch. Röntgenkongress, Wiesbaden 1964, G. Thieme, Stuttgart (1965) 57.
[3] SODEE, D., "Delineation of anatomic structures and the detection of carcinoma utilizing low energy mercury labelled chlormerodrin", Radioaktive Isotope in Klinik und Forschung 6 (1964) 167.
[4] EARLY, P.J., RAZZAK, M.A., SODEE, D.B., Textbook of Nuclear Medicine Technology, Mosby, St Louis (1969) 254.
[5] MUNDINGER, F., GERHARD, H., Distribution in the blood of radioisotopes used for cerebral tumour diagnoses studied in experimental and human brain tumours. Untersuchungen über die Verteilung der zur Hirntumordiagnostik verwendeten Radioisotope in der Blutbahn, in experimentellen Tumoren und menschlichen Hirngeschwülsten, Acta Neurochir. 11 (1963) 398.

---

[1] The examined preparations of ^{197}Hg-chlormerodrin and ^{197}HgCl$_2$ were prepared by the Commissariat à l'énergie atomique, Département des radioéléments, Saclay (France).

[6] DE ROO, M.J.K., Experimental evaluation of scanning agents for tumor localization, J. Nucl. Biol. Med. **16** (1972) 62.
[7] JOHNSTON, G.S., LARSON, A.L., McCURDY, H.W., Tumor localization in the nasopharynx using radiomercury labelled chlormerodrin, J. Nucl. Med. **6** (1965) 549.
[8] SODEE, D.B., Localization of eye tumors by external counting utilizing mercury ^{203}Neohydrin, J. Nucl. Med. **4** (1963) 194.
[9] WAGNER, H.N., Jr., Principles of Nuclear Medicine, Saunders, Philadelphia (1968) 664.
[10] FARRER, P.A., SAHA, G.B., A comparison of ^{111}InCl$_3$, ^{67}Ga and ^{197}HgCl$_2$ in tumor bearing mice, J. Nucl. Med. **15** (1974) 489.
[11] ISAAC, R., RAYNAUD, C., KELLERSHOHN, C., Détection scintigraphique des tumeurs malignes à l'aide du ^{197}HgCl$_2$, Nucl. Medizin **7** (1968) 97.
[12] RAYNAUD, C., ISAAC, R., Unpublished results.
[13] GARDEL, J., ISAAC, R., RAYNAUD, C., KELLERSHOHN, C., Détection scintigraphique des tumeurs malignes à l'aide du bichlorure de mercure (^{197}HgCl$_2$), Ann. Oto-Laryngo. (Paris) **84** (1967) 633.
[14] BUCHWALD, W., DIETHELM, L., HAAS, J.P., WOLF, R., "Ergebnisse Nuklearmedizinischer Untersuchungen beim Mammakarzinom", Radioisotope in der Lokalisationsdiagnostik, Schattauer-Verlag Stuttgart (1967) 549.
[15] WOLF, R., FISCHER, J., Tumorszintigraphie mit radioaktiven Quecksilberverbindungen (^{197}HgCl$_2$ und ^{197}Hg-Neohydrin). Radionuklide in der klinischen und experimentellen Onkologie, Zweite Jahrestagung der Gesellschaft für Nuklearmedizin, Heidelberg (1965) 223.
[16] RAYNAUD, C., ISAAC, R., RYMER, M., BLANCHON, P., MONOD, O., KELLERSHOHN, C., Scintigraphies au ^{197}HgCl$_2$, J. Fr. Med. Chir. Thorac. **21** (1967) 735.
[17] LAMY, P., BURG, C., ANTHOINE, D., VAILLANT, G., VAILLANT, D., MONNEAU, J.P., FROMENT, J., Intérêt de la scintigraphie au ^{197}HgCl$_2$ dans le diagnostic des images rondes intrapulmonaires, J. Fr. Med. Chir. Thorac. **23** (1969) 159.
[18] PECORINI, V., DEGROSSI, O., ARTAGAVEYTIA, D., CHWOJNIK, A., Diagnóstico de atipias utilizando bicloruro de mercurio radioactivo, Rev. Biol. Med. Nucl. **2** (1970) 197.
[19] ISAAC, R., RAYNAUD, C., RYMER, M., MONOD, O., Scintigraphies pulmonaires au ^{197}HgCl$_2$. Résultats cliniques, J. Fr. Med. Chir. Thorac. **3** (1970) 225.
[20] ISAAC, R., L'utilisation de la scintigraphie au bichlorure de mercure 197 pour le diagnostic des cancers thoraciques, Thèse de Médecine, Faculté de Médecine, Paris (1970).
[21] RESCIGNO, B., UGOLOTTI, G., MIGLIO, M., BOBBIO, P., La scintigraphia con bicloruro di mercurio radioattivo nella diagnostica polmonare, con particolare riferimento alle neoplasie, Riv. Patol Clin. Tuberculosi **44** (1971) 149.
[22] FARRER, P.A., SAHA, G.B., MUNRO, D.D., DOLLFUSS, R.E., MEAKINGS, J.F., MacLEAN, L.D., Radionuclide imaging of intrathoracic mass-lesions using ^{197}HgCl$_2$ and ^{99m}Tc macroaggregated human serum albumin, J. Nucl. Med. **15** (1974) 490.
[23] RAYNAUD, C., Généralités sur l'utilisation des substances radioactives pour l'étude des tumeurs pulmonaires, J. Fr. Med. Chir. Thorac. (in press).
[24] ISAAC, R., VAN QUAETHEM, M., RAYNAUD, C., BLANCHON, P., Scintigraphies pulmonaires au ^{197}HgCl$_2$ au cours d'atteintes néoplasiques et infectieuses du poumon, J. Fr. Med. Thorac. **7** (1969) 771.
[25] GOTTA, H., CHOWOJNIK, A., MARTINEZ SEEBER, J., PECORINI, V., Evaluation of pulmonary scintigraphy with mercurial compounds, Nucl. Med. **12** (1974) 275.
[26] RAYNAUD, C., The Renal Uptake of Radioactive Mercury, C.C. Thomas, Springfield, Ill. (in press).
[27] WAGNER, H.N., Jr. WEINER, I.M., McAFEE, J.G., MARTINEZ, J., 1-Mercuri-2-hydroxypropane (MHP), Arch. Intern. Med. **113** (1964) 696.
[28] BOWEN, H.J.M., Trace Elements in Biochemistry, Academic Press, London (1966) 116.

DISCUSSION

K. HISADA: We had some limited experience in 1967 with ^{203}Hg-haematoporphyrin sodium salts. We got good tumour images in cases of squamous cell cancer in the neck. The number of cases was limited as we couldn't get additional material, and the results were only published in a review on tumour scanning in Japanese. Hg-haematoporphyrin is a radiosensitizer as well as an anti-tumour agent. Does anyone know about other reports on this subject?

H.J. GLENN: I think the Japanese have done the most work with the haematoporphyrins. Many years ago we did prepare ^{67}Cu-haematoporphyrin, and I want to stress that many of the divalent compounds do form very firm bonds with haematoporphyrin: I want to use the word chelate. The only problem in the clinical trial was that the copper half-life was not long enough. We lost all the activity before it was shown to be of clinical value. There are some reports which state that if you fill in the "doughnut" in the middle of the haematoporphyrin molecule, you may change the biological distribution of those compounds.

W.H. BEIERWALTES: The group at Berkeley has done a considerable amount of work with haematoporphyrins, but I don't know if they have used the mercury label. I have a question for Dr. Comar: When you mentioned the uptake of mercury chloride in lung tumours particularly, and the lack of uptake in benign tumours I should like to ask you if you have studied the subcellular fractionation of the radiolabelled mercuric chloride, did it concentrate primarily in the mitochondrial or lysosomal fraction?

D. COMAR: Unfortunately I don't know of any publication on that. One technical problem that we encountered with mercury is that mercury is adsorbed everywhere. Usually 10 to 15% of the mercury stays in the syringe when you have a very high specific activity.

At the same time I should like to mention another isotope of mercury, that is $^{197}Hg^m$ which is cyclotron produced and has been prepared in Japan and by our group in Orsay. It can be prepared carrier-free, whereas ^{197}Hg, when used, will be injected in an amount of about a few micrograms. Another advantage of $^{197}Hg^m$ is its emission around 140 to 160 keV and for imaging of deep tumours in the lung it might be better than using the 80-keV rays of ^{197}Hg. One difficulty, of course, is that it decays into ^{197}Hg, the half-life of $^{197}Hg^m$ being 24 hours and that of ^{197}Hg 64 hours.

H.J. GLENN: Dr. Comar, can you tell me if the specific activity of ^{197}Hg-mercuri-hydroxypropane (or bromo-mercuri-hydroxypropane) is important for a successful spleen scan? One of the things that interfered with the commercial marketing of this compound in the United States was that it was thought necessary to furnish an injectable, non-radioactive material as well. This threw the compound out of the investigative drug category of radioisotopes in the United States and put the non-radioactive material into the standard drug category, and thereby created a great difference as to how much effort was thought necessary to get regulatory approval.

H. LANGHAMMER: Does it make any difference in the accumulation of mercury compounds in tumour tissues whether the patients have been treated or not?

D. COMAR: Yes it does, and that is the interesting point since you can follow the progress of the treatment. Patients treated by radiotherapy as well as by chemotherapy show very low or no uptake of ^{197}Hg.

V.R. McCREADY: Have you compared the uptake of mercury, gallium and bleomycin in specimens, and is there a higher concentration of mercury?

D. COMAR: We have done this in animals and in humans. In mice we compared bleomycin, mercury, cobalt, gallium and a lot of other isotopes. It seems as if each compound gives about the same ratio to the surrounding tissues. We only looked for the uptake 24 hours after injection, but we couldn't find any significant difference between the different nuclides. In humans it seems that the tumour uptake is slightly higher with bleomycin. The gallium uptake is the same as the mercury uptake.

H.J. GLENN: How does the mercuric chloride technique compare with the X-ray techniques for localizing lung tumours?

D. COMAR: Malignant tumours can be detected earlier with ^{197}Hg than with X-rays. This method is used routinely for the detection of relapses in the follow-up of operated lung cancer. The most difficult problem is that inflammatory lesions also will give positive results.

R.M. KNISELEY: I gather that the incidence of inflammatory lesions is high enough, so I remain sceptical of its value as a tool that will allow the physician or surgeon to make a decision on managing the patient as to benign or malignant. The same seems to be true of gallium, selenite, indium and the other cation-type compounds.

D. COMAR: At the present time no isotope technique is able to differentiate malignant tumours from inflammatory lesions.

R.M. KNISELEY: You have not only benign lesions which do pick up mercury, but you also have malignant lesions which don't concentrate gallium or mercury.

V.R. McCREADY: I think we have to define clearly the purpose of our tumour-localizing studies. They may be used for cancer screening or their prime purpose may be to distinguish between benign and malignant lesions. In combination with other negative signs even a negative scan may confirm a clinical decision that treatment is not needed. Then we have the situation where we want to know if there are other areas with tumour which haven't been found by radiological methods in a patient with proven cancer, or we may simply want additional information, for instance before surgery.

K. HISADA: To me, too, there are some definite indications for tumour scintigraphy: (1) The differential diagnosis in a patient with a mass lesion suspected of malignancy. This is, of course, difficult because the present tumour-localizing agents also accumulate in benign lesions, but it may be important to follow the patient and fate of the radioactive accumulation. (2) The search for metastatic lesions in patients with known primary focus. (3) The early detection of recurrent carcinoma and the subsequent management of the patient. (4) The staging of tumours, especially Hodgkin's disease. (5) In order to set up the radiation fields, especially in the mediastinal and pelvic areas; and (6) To follow the response of the lesion to non-surgical therapy such as radiotherapy or chemotherapy.

V.R. McCREADY: Currently there are several tests, e.g. CEA, which may detect occult cancer by examining blood or urine. If successful, there will be a need for an efficient way of localizing the position of the tumour in the patient. Whole-body scintigraphy is potentially the simplest way of doing this.

W.H. BEIERWALTES: I do think we have cases where we can differentiate benign from malignant tumours. The iodochloroquine analogue is taken up by the ocular melanoma and also by the dermal melanoma, whereas diseases which are not malignant don't show any accumulation of the chloroquine analogue. Similarly, you have some high risk groups where you can prove the presence of malignant lesions, for instance in patients with proved carcinoma of the thyroid, where you may find lung metastases with a negative chest X-ray, or patients or relatives of a person with a medullary carcinoma where you can find the disease in about half of his relatives, even if the disease hasn't produced signs or symptoms.

E.H. BELCHER: Is there any information on differences between the dynamic patterns of uptake and clearance of mercury in malignant and benign lesions?

D. COMAR: There is no difference in kinetics of the uptake and release of mercury, unfortunately. I think this holds for quite a number of the other compounds which we use. The only thing that might be of interest in this connection is radioactive copper. It seems that the rates of uptake and release of carrier-free copper in tumours are different from those in inflammatory lesions. So far we have studied some ten to fifteen patients and in all cases we have been able to differentiate malignant tumours from inflammatory or benign tissue.

V.R. McCREADY: I think we should stress the localization more than the differential diagnoses in our present situation.

W.H. BEIERWALTES: I think that Dr. McCready's generalization is generally correct. But we do have cases where we can give the correct differential diagnosis, for instance with the combined ultrasound probe and the ^{125}I chloroquine analogue. This is very important because we may save eyeballs which would otherwise be ablated.

V.R. McCREADY: In the cases in which you have made a differential diagnosis, you have been in a special situation where the tumours were hormone producing, with one exception, the melanoma, which has a very specific feature that you have just described. My previous remarks were related to tumours which are non-hormone producing.

IAEA-MG-50/21

CLINICAL AND EXPERIMENTAL STUDIES OF SELENIUM-75-LABELLED COMPOUNDS

A review

A.H.G. PATERSON, V.R. McCREADY
Department of Nuclear Medicine,
Royal Marsden Hospital,
Sutton, Surrey,
United Kingdom

Abstract

CLINICAL AND EXPERIMENTAL STUDIES OF SELENIUM-75-LABELLED COMPOUNDS: A REVIEW.
    Clinical experience with ^{75}Se-selenite and ^{75}Se-selenomethionine is reviewed. Reports of improved tumour specificity with ^{75}Se-selenite compared with other tumour-imaging agents are examined and the value of scanning certain tumours with ^{75}Se-selenomethionine is assessed. Its value in the differential diagnosis of liver disease is also discussed. The study of the mechanism of uptake of these compounds provides a comparison with studies of other tumour-imaging agents. Clinical studies have suggested that ^{75}Se-selenomethionine uptake into lesions is related to the vascularity of the lesion. Experimental work involving the measurement of ^{75}Se incorporation into the protein of tumours has suggested that increased rates of tumour protein synthesis are a contributing factor in the accumulation of ^{75}Se. Tumour imaging by means of ^{75}Se-labelled compounds is compared and contrasted with tumour imaging by means of ^{67}Ga-citrate.

    Selenium-75-labelled compounds have been used since the mid-sixties for the demonstration of certain tumours by the technique of whole-body scanning, although following the introduction of ^{67}Ga-citrate in 1969 their use for this purpose has diminished considerably. However, they remain an interesting group of compounds since reported clinical and experimental studies of their uptake into tumours provide an interesting comparison with studies of the uptake mechanisms of the more recently introduced tumour-imaging agents.
    The first ^{75}Se compound to be used for tumour detection was ^{75}Se-selenite. Its introduction for this purpose was initially based on the hypothesis that because of similarities in the chemical properties of selenium and sulphur, tumours which laid down chondroitin sulphate (i.e. cartilaginous tumours) might be detected by incorporating a suitably labelled selenium compound. This, in fact, turned out to be correct (Esteban et al. [1]), but it was not realized at that time that the phenomenon of nuclide uptake into tumours was much more generalized than merely uptake into cartilaginous tumours. It was not until later when Cavalieri et al. [2], and Herrera et al. [3] more or less simultaneously reported their observations on ^{75}Se-selenite and ^{75}Se-selenomethionine respectively that this became apparent.
    The former group based their clinical study on observations made in sarcoma-bearing rats and they found that a variety of tumours could be detected using ^{75}Se-selenite. These tumours included cerebral astrocytomas, bronchogenic carcinomas, a case of myelomatosis and a case of carcinoma of the colon. These investigators felt that ^{75}Se-selenite was more tumour specific than any agent previously studied, since cases of cerebral infarction, lung abscess and pneumonia which they examined showed no significant uptake on scanning. Baptista [4] has since studied 58 patients with a variety of hepatic lesions by comparative scintigraphy using ^{198}Au- or ^{99}Tc^{m}-colloid and ^{75}Se-selenite. He has supported the suggestion that ^{75}Se-selenite is more tumour

specific than other agents. Of 15 patients with benign lesions of the liver (e.g. abscesses, cysts, cirrhosis) all gave negative scans, whereas the remaining 43 patients with primary or metastatic liver disease all exhibited positive uptake of ^{75}Se. Esteban et al. [5] reported very encouraging but less specific results with ^{75}Se-selenite for lesions in the same area. However, there has been little further confirmatory work reported in the literature on this aspect of ^{75}Se-selenite metabolism in other disease areas, since most subsequent studies have used ^{75}Se-selenomethionine.

While pancreas scanning a patient with ^{75}Se-selenomethionine, Herrera et al. [3] observed heavy uptake in a retro-peritoneal mass of lymphosarcoma. These workers then went on to examine a further nine patients with a variety of lymphomas and recorded a gratifying success rate. They also observed that radiotherapy very quickly caused the increased uptake to return to normal. However, again, it was not appreciated that ^{75}Se-selenomethionine was taken up by a wider range of tumours than the lymphomas they had examined, and it was not until further studies were published that this became apparent. D'Angio et al. [6] showed that neuroblastoma in children could be scanned successfully and hepatomas have also been recorded as exhibiting good ^{75}Se-selenomethionine uptake [7]. Kaplan and Domingo [8] found that the diagnostic value of liver scanning could be enhanced by using a dual-channel subtraction technique with ^{99}Tc^m-sulphur-colloid and ^{75}Se-selenomethionine. They correlated ^{75}Se uptake with the vascularity of various types of lesions as demonstrated by hepatic angiography. Thus, the relatively avascular amoebic abscesses and metastases from tumours of the bronchus or pancreas showed little uptake, but the highly vascular hepatoma invariably showed good uptake. It is interesting to note that these results are broadly similar to those obtained by Lomas et al. [9] and Suzuki et al. [10] who used ^{67}Ga-citrate to demonstrate hepatic lesions. It must be added in parentheses here that the importance of isotopic methods in the differential diagnosis of liver disease is likely to decline somewhat with the development of grey-scale ultrasonic scanning of this area. Thomas et al. [11] showed that uptake occurred not only in malignant thyroid lesions but also, to a lesser extent, in benign thyroid lesions.

In a study designed to assess the practical value of scanning with ^{75}Se-selenomethionine in lymphomas and seminomas, Ferruci et al. [12] compared the abdominal findings in patients with lymphoma by using X-ray lymphangiography and isotope scanning. In fact, the ^{75}Se-selenomethionine method compared quite favourably with X-ray lymphangiography although in one-third of positive lymphangiograms the scan was negative. Conversely, however, the scan was positive in 25% of cases where the X-ray lymphangiogram was difficult to interpret or reported as normal. Some of these may have been false positives but the majority indicated disease not initially detectable on the X-ray. One especially interesting point in this study was the report of extremely good results in the scanning of seminoma of the testis; this is a tumour we ourselves have found to be very satisfactorily scanned with ^{67}Ga-citrate [13].

Studies of the mechanism of uptake of ^{75}Se-selenomethionine provides interesting reading. As mentioned already, the clinical studies on hepatic lesions seemed to indicate some relationship between uptake of ^{75}Se-selenomethionine and tumour vascularity. This aspect has not been adequately examined in animals and must remain an open question. However, the striking similarities in the nuclide uptake characteristics of various lesions between ^{75}Se-selenomethionine and ^{67}Ga-citrate argues strongly in favour of the suggestion that tumour blood vessel permeability may play an important role in the uptake of many agents into tumours.

Since ^{75}Se-selenomethionine is a labelled essential amino acid, it is not unreasonable to postulate that an increased rate of protein synthesis within tumours is a cause of its concentration within tumours, and the contribution of tumour protein synthesis to the uptake of ^{75}Se-selenomethionine has indeed been studied in some detail. It has been well shown that certain tumours, among which are included lymphomas and seminomas, have a more rapid rate of protein turnover than have normal tissues [14]. Spencer et al. [15] studied the incorporation of ^{75}Se-selenomethionine into tumour proteins (T.C.A. insoluble fractions) in lymphoma-bearing mice over a period of time. They found that the incorporation of ^{75}Se into protein increased steadily with time in both tumour and liver reaching 40% of the total tumour and liver ^{75}Se at one hour after injection. The concomitant administration of inhibitors of protein synthesis (actinomycin D or puromycin) lowered the amount of ^{75}Se which was protein bound, although whether or not the total tumour uptake was affected is not mentioned.

Although the biochemical and nutritional aspects of selenium metabolism have been the subject of considerable study, mechanisms of uptake into tumours other than the possible relationship to protein metabolism outlined here have not been looked at to any great extent. Selenite is a potent oxidizer of sulphydril groups [16], and this is a reason for its rapid fixation in normal tissues, and may account for its accumulation in tumours as well. Another suggested possibility [17] is that selenite is metabolized by intestinal flora to selenomethionine in which form it might then enter the tumour.

In the last few years other agents such as ^{67}Ga-citrate and labelled bleomycin compounds have taken over from ^{75}Se-labelled compounds as the tumour-imaging agents of choice. There have been no direct comparative studies of these agents with ^{75}Se-selenomethionine, although the clinical impression is that the more recent agents have more favourable tumour-to-background radioactivity ratios. The greatest disadvantage of the ^{75}Se-labelled compounds is their long biological half-life which for ^{75}Se-selenomethionine is 45 days and for ^{75}Se-selenite 65 days.

In conclusion, we would like to draw attention to some of the striking clinical similarities in the medical conditions producing sufficient uptake to give a positive scintiscan with ^{75}Se-selenomethionine and ^{67}Ga-citrate. These similarities inevitably lead one to postulate that there may be similarities in mechanisms of uptake of some of these compounds and that these biological phenomena may indeed be the most important factor in our ability to detect tumours. That there are secondary mechanisms occurring is well illustrated by the experimental studies described here, but whether these mechanisms are the most important from the point of view of tumour imaging remains to be seen.

REFERENCES

[1] ESTEBAN, J., LASA, D., PEREZ-MODREGO, S., Detection of cartilaginous tumours with selenium-75, Radiology 85 (1965) 149.
[2] CAVALIERI, R.R., SCOTT, K.G., SAIRENJI, F., Selenite (^{75}Se) as a tumour localising agent in man, J. Nucl. Med. 7 (1966) 197.
[3] HERRERA, N.E., GONZALES, R., SCHWARTZ, R.D., DIGGS, A.M., Belshy, J., ^{75}Se-methionine as a diagnostic agent in lymphoma, J. Nucl. Med. 6 (1966) 792.
[4] BAPTISTA, A.M., "Positive gammagraphy of tumours with ^{75}Se-selenite: distinction of cancer from benign hepatic tumours", Medical Radioisotope Scintigraphy 1972 (Proc. Symp. Monte Carlo, 1972) 2, IAEA, Vienna (1972) 701.
[5] ESTEBAN, J., VAZQUEZ, R., FOMBELLIDA, J.C., CABALLERO, O., LLOPIS, A., "Positive diagnosis of tumours with selenium-75 selenite", Medical Radioisotopes Scintigraphy 1972 (Proc. Symp. Monte Carlo, 1972) 2, IAEA, Vienna (1972) 687.
[6] D'ANGIO, G.T., LOKEN, M., NESBIT, M., Radionuclear (^{75}Se) identification of tumour in children with neuroblastoma, Radiology 93 (1969) 615.
[7] STOLZENBERG, J., Uptake of ^{75}Se-selenomethionine by hepatoma, J. Nucl. Med. 13 7 (1972) 565.
[8] KAPLAN, E., DOMINGO, M., ^{75}Se-selenomethionine in hepatic focal lesions, Semin. Nucl. Med. 2 (1972) 139.
[9] LOMAS, F., DIBOS, P.E., WAGNER, H.N., Jr., Increased specificity of liver scanning with the use of ^{67}gallium citrate, New Engl. J. Med. 286 (1972) 1323.
[10] SUZUKI, T., HONJO, I., HAMAMOTO, K., Positive scintiphotography of cancer of the liver with ^{67}Ga-citrate, Am. J. Roengenol., Radium Ther. Nucl. Med. 113 (1971) 92.
[11] THOMAS, C.G., PEPPER, F.D., OWEN, J., Differentiation of malignant from benign lesions of the thyroid using complementary scanning with ^{75}Se-selenomethionine, Ann. Surg. 170 (1969) 396.
[12] FERRUCCI, J.T., BERKE, R.A., POTSAID, M.S., ^{75}Se-selenomethionine isotope lymphography in lymphoma: Correlation with lymphangiography, Am. J. Roentgenol., Radium Ther. Nucl. Med. 109 (1970) 793.
[13] PATERSON, A.H.G., PECKHAM, M.J., McCREADY, V.R., Value of gallium scanning in seminoma of the testis, Br. Med. J. 1 (1976) 1118.

[14] QUASTEL, J.H., BICKIS, I.J., Metabolism of normal tissues and neoplasms, Nature (London) **183** (1959) 281.
[15] SPENCER, R.P., MONTANA, G., SCANLON, G.T., EVANS, O.R., Uptake of selenomethionine by mouse and in human lymphomas, with observations on selenite and selenate, J. Nucl. Med. **8** (1967) 197.
[16] PAINTER, E.P., The chemistry and toxicity of selenium compounds, Chem. Rev. **28** (1941) 179.
[17] SHRIFT, A., Biochemical inter-relations between selenium and sulphur in plants and micro-organisms, Fed. Proc. **20** (1961) 695.

DISCUSSION

R.L. HAYES: At my institution we have used ^{75}Se-selenomethionine for tumour studies in animals and we found a considerably lower concentration of this compound in tumour than of other tumour-localizing agents. But to some extent it depends on the nature of the tumour.

V.R. McCREADY: There are several reasons for not being too enthusiastic about ^{75}Se-selenomethionine. You can't give more than 250 μCi, and it is a poor radionuclide for imaging with a gamma-camera. 250 μCi will give you a whole-body dose of about 1.5 rad, 10 rads to the kidney and 7 rads to the liver. In our hospital we carry out many renograms and many follow-up scans, and when so, it is unacceptable to have a radioisotope that remains in the body for a long time. Finally, selenomethionine is expensive; we have to pay about US$ 40 per study. From the theoretical point of view, selenium-labelled methionine is an unnatural substance which should not be introduced into human metabolism.

R.L. HAYES: Selenium can cause a rather severe debilitation of cattle when eating forage with a high selenium content.

V.R. McCREADY: One basic reason for using ^{75}Se-selenomethionine may be to study the incorporation of proteins in the tumours. But there is very little on the basic physiology and biochemistry in the literature. As to pancreatic scanning, you can summarize that if the pancreas scan appears to be normal, then the organ will be normal in about 80% of the cases. If it has a non-radioactive area, nearly always there will be something wrong with the patient. The actual abnormality may vary from a hiatus hernia to an aortic aneurysm, or even tumour or pancreatitis. Subtraction scanning may help to clarify the situation, and it is easy to do. But it is reasonable to mention that we have completely missed some lesions where there was a very obvious pancreatic tumour at laparotomy. We have tried to explain this by the rapid growth of the tumour incorporating a lot of selenomethionine, resulting in an area just as radioactive as the normal pancreatic tissue.

D. COMAR: When you mention the drawbacks of the long half-life of ^{75}Se, do you think it would be worth while to use ^{72}Se, which has an 8-day half-life, but unfortunately an energy as low as 46 keV in 69% of the disintegrations? It doesn't seem easy to produce, and it decays into radioactive arsenic. But ^{72}Se can be produced in a carrier-free state.

V.R. McCREADY: The important point is whether incorporation of amino acids into tumour will reflect any specific aspect of tumour metabolism. But we need better count-rates, say, for instance 2 mCi instead of the present 200 μCi. The actual concentration ratios are rather poor, but still if the incorporation of methionine could be a useful indicator of some important clinical aspect of the tumours, for instance as an indicator of the response to chemotherapy or to radiotherapy, then it would be worth while following this line.

T. MUNKNER: Labelled selenomethionine was originally produced by biosynthesis; today it is also available with a much higher specific activity, made by chemical synthesis. Do you have any comments on the differences between these two compounds?

H.J. GLENN: From the point of synthesis, it will be an advantage if we could use selenium in a carrier-free form. Biochemical synthesis will only work up to a certain concentration of selenium and after that your yeast begins getting poisoned. A couple of years ago, there was a lot of discussion about the differences between material made by biosynthesis and that made by chemical synthesis. The biosynthesis gives you the L-form almost exclusively, and the chemical synthesis gives you the DL-form. I wonder if someone in the panel could tell me about possible consequences of this difference. Finally, I would like to mention that there was a tremendous discussion several years ago among some scientists about whether or not trace amounts of selenium in the diet were actually the cause of certain cancers or whether they were the cure of mechanisms involved in it.

R.L. HAYES: Before selenomethionine was introduced as a localizing agent, some 15 or 20 other amino acids labelled with ^{14}C were studied rather extensively, and it was shown that tryptophane was by far the best agent from the standpoint of pancreas-to-liver ratio. None of the amino acids showed any particular preference for tumour tissue, that is the tumour-to-liver ratio was around one or possibly up to two. Blau's attempt to utilize an iodine-labelled tryptophane didn't bear results and that was the reason why he went on to selenomethionine, not because this was the amino acid or the agent of choice. We have ourselves been particularly interested in ^{11}C-labelled tryptophane with respect to visualization of normal pancreas.

W.H. BEIERWALTES: The uptake of amino acids in the cells is a very complicated process. We realized this when we labelled phenylalanine with iodine, in the ortho-, meta-, and para-positions, respectively. Each of these compounds was concentrated with species specificity in the mouse pancreas. The uptake of the amino acid followed in two phases: an immediate one, which may be related to the transport across the cell membrane, and a delayed phase, which perhaps is due to the incorporation into peptides. When we went on to study the same compounds in the dog and the the monkey we didn't find any uptake of these compounds.

R.L. HAYES: The point of species specificity is very interesting and seems to be a pitfall that keeps cropping up. This was also the experience with L-amino-cyclopentane-carboxylic acid, which has a phenomenal deposition in rat pancrease as well as in tumour tissues, but in further studies the pancreas localization was shown to be species specific.

V.R. McCREADY: It seems to me as if we are missing a top specialist in protein chemistry who could advise us on the best methods for using labelled amino acids for protein synthesis. Differences in protein synthesis might be a sensitive tumour measure, for instance following therapy. If we could have proteins labelled with ^{123}I we might have much better tumour-localizing agents than the present ones.

H.J. GLENN: In our work with para-iodophenylalanine there was a difference even between mice and rats in their liver:pancreas ratios. It seems as if the iodine molecule still permits the para-iodophenylalanine to accumulate in the pancreas but not to participate in amino acid synthesis, whereas the ^{14}C-labelled compound is not only accumulated but is also used in the subsequent amino acid or peptide synthesis.

TUMOUR SCINTIGRAPHY WITH GALLIUM-67

Present status

H. LANGHAMMER, G. HÖR, K. KEMPKEN, H.W. PABST,
P. HEIDENREICH, H. KRIEGEL
Nuclear Medicine Clinic and Polyclinic,
Technical University of Munich
and
Department of Nuclear Biology,
Society of Radiation and Environmental Research (GSUF),
Neuherberg/Munich,
Federal Republic of Germany

Abstract

TUMOUR SCINTIGRAPHY WITH GALLIUM-67: PRESENT STATUS.
The paper reviews the production and physical properties of ^{67}Ga and the biological fate of ^{67}Ga-citrate. Special emphasis is laid on the diagnostic use and limitations of ^{67}Ga for tumour scintigraphy. On the basis of the authors' clinical and experimental data as well as on results in the literature, it is concluded that the diagnostic value of ^{67}Ga-citrate as a tumour-localizing agent is limited by: (1) the non-specific uptake of ^{67}Ga, i.e. its lack of tumour specificity, and (2) the fact that the sensitivity of tumour detection depends on the percentage uptake, the tumour-to-background ratio, tumour size and type, and the viability of the tumour cells.

The introduction of ^{67}Ga-citrate by Edwards and Hayes [7] resulted in the widespread clinical use of this soft-tissue tumour-detecting agent in nuclear medicine, and ^{67}Ga has attracted much attention in clinical research in recent years.

Compared with other successful radiopharmaceuticals like ^{131}I-albumin, ^{75}Se-selenomethionine, and ^{197}Hg- and ^{203}Hg-chlormerodrin, ^{67}Ga is superior in absolute tumour concentration and in the ratio of tumour to normal tissue concentration [8, 19]. But the increased uptake of ^{67}Ga and of other tumour-seeking radiopharmaceuticals in a variety of non-neoplastic affections is a limitation in tumour scintigraphy. ^{99}Tc^{m}-labelled bleomycin has recently been reported to be superior to ^{67}Ga because of its higher detectability in adeno-carcinomas and lower affinity for inflammatory and granulomatous lesions [46].

PRODUCTION AND PROPERTIES OF GALLIUM-67

Gallium-67 is produced by irradiation of natural zinc with protons in a cyclotron [31]. At the same time large amounts of short-lived ^{66}Ga and ^{65}Ga are formed. Five days after bombardment the final product[1] contains no ^{65}Ga, less than 0.5% ^{66}Ga, and less than 0.04% ^{65}Zn per 1 mCi ^{67}Ga. The amount of copper is less than 5 μg/ml. These data apply to the solution that is injected; 1 ml of it (pH 6 − 8) contains 1 mCi ^{67}Ga, 1.6−2.5 mg sodium citrate, and 9 mg benzyl alcohol in isotonic

---

[1] N.V. Philips-Duphar, Cyclotron and Isotope Laboratories, Petten, The Netherlands.

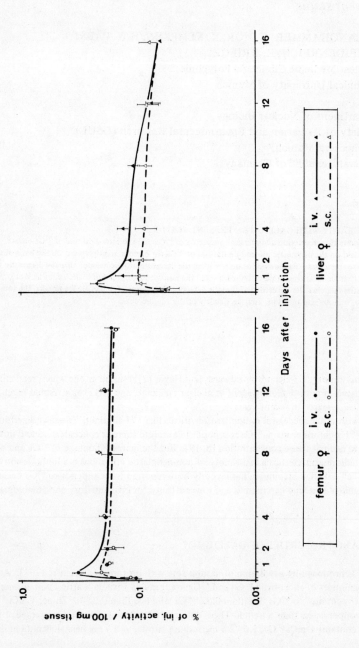

FIG.1. ^{67}Ga distribution in rats.

solution of sodium chloride. ^{67}Ga has a physical half-life of 78 hours and decays by electron capture followed by the emission of four main gamma rays of 92 to 388 keV.

Main energy (keV)	Output (%)
92	39.6
185	22.9
296	20.5
388	8

METHODS AND TECHNICAL PROCEDURE

A dose of 1.5–2.5 mCi of ^{67}Ga, usually in citrate form, was administered intravenously. Scanning was performed regularly 48 (to 72 hours) after injection using a rectilinear scanner equipped with a 5-in (12.5-mm) NaI crystal. For visualizing lesions near the surface a fine-focusing collimator was used, and for deeply situated lesions a coarse-focusing collimator was used by us and others [13, 14, 15, 17, 20, 27, 30, 40, 41]. According to the technical specifications of the scanner, measurements were performed differentially (296 keV gamma transition) or integrally (lower window at 250 keV).

Other authors have used a scintillation camera [16, 56, 57, 62].

KINETICS AND RADIATION DOSE

In animal studies on the distribution of radiogallium (Fig. 1), the highest concentration of ^{67}Ga was observed in femur (0.2% of dose/100 mg), followed by liver, spleen and kidney [28, 30]. The activity concentration curve in blood was interpretable as bi-exponential with a biological half-life of about 7 hours for the rapid component and 6.5 days for the slower one. Excretion of ^{67}Ga amounted to 29% of the injected dose 24 hours after injection and 50% after 3 days. Predominantly renal clearance was observed in the first 12 hours after injection; faecal excretion was greater in the following period of observation.

Special attention should be devoted to the finding that during breast feeding ^{67}Ga is accumulated progressively in neonatals, more than is for instance ^{99}Tc^{m} (Fig. 2). Mammary glands of lactating rats contain a 4 times higher concentration of ^{67}Ga than do the mammary glands of non-lactating animals (Fig. 3). This means that ^{67}Ga should not be used during the lactating period.

The radiation dose of ^{67}Ga depends on its cumulated concentration of activity. In experimental and clinical investigations preferential uptake was seen in skeleton, liver, spleen and kidneys [11, 19, 47, 50, 53]. The effective half-life ranged from 53 to 74 hours due to the relatively short physical half-life of ^{67}Ga, whereas the biological half-life of ^{67}Ga differed in various organs in the range from 162 to 850 hours [11, 53, 41]. On the basis of calculations for several target organs in man, according to the MIRD publications [45] the absorbed dose amounts to 250 mrads/mCi of ^{67}Ga for the total body if a non-homogeneous distribution of ^{67}Ga is assumed [11, 41]. The absorbed dose is highest in the liver (1330 mrads/mCi of ^{67}Ga).

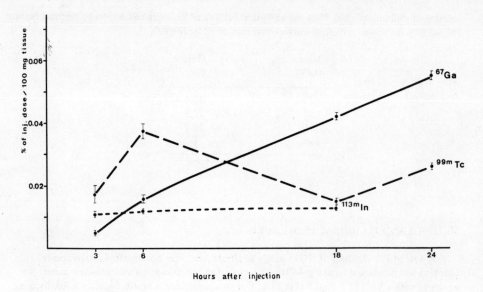

FIG.2. Retention of ^{67}Ga, $^{99}Tc^m$ and $^{113}In^m$ in mammary glands during lactation.

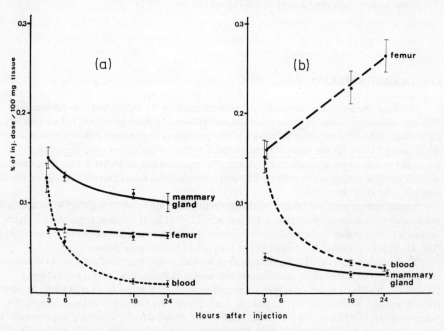

FIG.3. Distribution of ^{67}Ga.
(a) In lactating rats
(b) In non-lactating rats.

DIAGNOSTIC VALUE OF GALLIUM-67 IN TUMOUR SCINTIGRAPHY

To outline the clinical value of ^{67}Ga based upon clinical and experimental data, the following points should be considered:

1. Non-specific uptake of gallium-67

The mechanism of ^{67}Ga accumulation in tumours is still debated [1, 17, 26, 58] although, according to the present state of knowledge, it may be assumed that lysosomal association of ^{67}Ga is one of the predominant factors in the uptake of ^{67}Ga [26, 58].

Gallium-67 shares with all other tumour-seeking radiopharmaceuticals the disadvantage of being accumulated in non-neoplastic tumours as well as in various inflammatory diseases, including abscesses. Benign granulomas, particularly sarcoidosis, show an increased uptake of ^{67}Ga [9, 15, 20, 25, 27, 38, 41, 44]. The increased pulmonary concentration of ^{67}Ga in pneumoconioses is also noteworthy, especially in silicosis and asbestosis [13, 54]. In addition, ^{67}Ga has been used for the detection of myocardial infarcts [39]. Because of its non-specific uptake histological verification is necessary to establish definitely the tumour diagnosis.

2. Diagnostic accuracy of gallium-67

The detectability of tumours by means of ^{67}Ga scintigraphy depends on the size and depth of the tumour and on the localization within the body. Tumour processes near the surface (with a diameter of 0.8 cm) are detectable, but those remote from the surface are not visualized below 2.0 cm [10, 15, 25]. The scintigraphic detectability depends on the tumour localization. It is remarkable that 80–90% of malignant processes can be delineated outside the abdominal area, whereas intra-abdominal tumours are often not distinguishable from ^{67}Ga in the bowels and liver.

An analysis of scintigrams performed at various institutions in 1966 leads to the following results (Table I). Positive scans were obtained in 73.7% of all the cases, better results being obtained in the subcategories of malignant lymphomas and intrathoracically situated neoplasms. This high percentage of positive scans, 86.9%, explains the accuracy of tumour detection within the chest, particularly in bronchial carcinomas.

Gallium-67 scanning is especially valuable when disclosing inside and outside the thorax additional foci which were not revealed by X-ray or other clinical procedures [15, 19, 32, 33, 38, 40]. In 151 untreated patients with Hodgkin's disease, Johnston et al. [33] demonstrated one or more positive sites in approximately 90% of their scans.

Gallium-67 has also been used in the detection of tumours in the liver [16, 20, 43, 56, 57]. Larson et al. [43] found ^{67}Ga to be more sensitive than radiocolloids as an indicator of hepatic involvement in Hodgkin's disease. In contrast to this result, Suzuki et al. [57] have made the following observations in 39 patients with metastatic carcinomas: 27 cases with decreased, 10 cases with similar and 2 cases with slightly increased ^{67}Ga uptake, whereas hepatomas generally had an increased accumulation of radiogallium. It may be concluded that ^{67}Ga is only valuable for the identification of hepatomas, which cannot be detected with the AFP test, and for differentiating them from other AFP-negative lesions such as cholangioma, cirrhosis and metastatic tumours of the liver. The ^{67}Ga scan, however, is not a reliable tool in the diagnosis of metastatic involvement of the liver.

Gallium-67 accumulation in malignant thyroid tumours has been reported by numerous investigators [7, 8, 9, 25, 29, 34, 36, 41] and their results indicate that thyroid carcinoma is not in all cases unequivocally diagnosable with radiogallium. Our own experience deals with 21 patients with various histological types of thyroid carcinoma. 17 cases had an undoubtedly positive scan, without any correlation with the differences in morphology. On the basis of a review of the

TABLE I. RESULTS OF GALLIUM-67 SCINTIGRAPHY IN TUMOUR DETECTION

References	Cases Pos./total	Lymphoma Pos./total	Thoracic cavity Pos./total
Berelowitz and Blake [2]	29/63	6/8	1/1
Cellerino et al. [3]	34/36		34/36
Dvorak and Moritz [5]	90/144	12/12	35/53
Fogh and Edeling [9]	118/135	5/7	70/73
Fröhlich et al. [10]	96/105		75/84
Grebe et al. [15]	57/83		35/38
Hayes and Edwards [19]	74/140	33/51	11/20
Heidenreich et al. [20]	19/32	6/12	3/4
Higasi and Nakayama [25]	51/79	2/2	25/35
Hör et al. [27]	121/194	18/31	56/63
Ito et al. [32]	32/32	7/7	21/21
Johnson et al. [33]	137/151	137/151	
Kay and McCready [35]	76/96	76/96	
Kinoshita et al. [38]	102/95		103/95
Langhammer et al. [40]	137/203	22/35	64/72
Langhammer et al. [41]	96/120	36/51	67/86
Lavender et al. [44]	8/24	1/2	
Ramos et al. [51]	86/120		12/19
Riccabona et al. [52]	54/64		
Turner et al. [59]	11/12	11/12	
Vaidya et al. [60]	11/20	2/3	0/2
Winchell et al. [62]	11/18	5/5	0/1
Total cases	1450/1966	379/485	611/703
Per cent positive	73.7%	78.1%	86.9%

literature and their own results on thyroid malignancies, Kaplan et al. [34] have recently pointed out that ^{67}Ga has only limited application in the diagnosis of solitary cold nodules, although it may be of value in identifying anaplastic thyroid tumours.

Gallium scanning has been applied with general success for staging in systemic malignancies of RES and other malignant lymphomas, particularly in Hodgkin's disease [19, 23, 33, 35, 59]. The recent report by Johnston et al. [33] revealed that approximately 90% of the scans demonstrated at least one positive site in untreated patients with Hodgkin's disease and that 73% of the histologically proven or clinically evident tumour sites were positive. A "false positive" rate of less than 5% was also remarkable. However, ^{67}Ga scanning should not be relied upon alone in staging because of its low level of reliability for abdominal lesions and because a negative scan does not definitely rule out the possibility of disease.

TABLE II. GALLIUM-67 CONCENTRATION IN SAMPLES OF HUMAN TISSUE (n)

Morphology	n	% inj. dose/100 mg wet tissue $\times 10^{-4}$ ($\bar{x} \pm$ SD)		
Normal tissues				
Lung	34	3.197	±	1.577
Bronchus	26	4.451	±	2.008
Pleura	9	3.692	±	2.181
Spleen	12	10.685	±	5.623
Liver	8	11.078	±	5.044
Inflammation				
Pneumonia (chronic)	15	7.894	±	4.084
Bronchitis	6	9.114	±	3.265
Connective tissue	21	3.266	±	1.554
Necrosis	17	1.782	±	1.751
Necrosis (up to 30% tumour tissue)	6	7.969	±	2.415
Tuberculosis	5	11.672	±	4.207
Tumours				
Differentiated carcinoma				
Bronchiolo-alveolar cancer	13	18.002	±	5.362
Epidermoid cancer (keratinizing)	14	21.082	±	4.709
Epidermoid cancer (moderately differentiated)	12	19.729	±	9.072
Undifferentiated carcinoma				
Solid cancer (partly glandular differentiated)	4	22.939	±	10.327
Solid cancer (undifferentiated)	7	19.120	±	5.505
Small cell anaplastic cancer	7	22.751	±	5.626
Giant cell cancer	3	45.778	±	22.431
Adeno-carcinomas				
Stomach	5	21.947	±	6.829
Metastasis of a colon cancer				
Adenomatoid hamartoma	5	12.800	±	3.750
Various tissues				
Muscle	2	4.232		
Oesophagus mucosa	1	1.281		
Mucosa of fundus	1	3.704		
Lymph node (without tumour)	25	14.771	±	10.480
Aspergillosis of lung	2	2.002		
Lymphogranuloma	3	35.591	±	17.830

TABLE III. SUMMARY OF RESULTS OF TUMOUR SCINTIGRAPHY BY MEANS OF GALLIUM-67 IN RELATION TO PREVIOUS THERAPY

Total number of cases	Untreated		Treated		References
	Positive	Negative	Positive	Negative	
151	137	14	–	–	Johnston et al. [33]
103	55	15	8	25	Haubold and Aulbert [17]
99	87	1	6	5	Kinoshita et al. [38]
120	78	–	17	25	Langhammer et al. [41]
203	121	43	16	23	Langhammer et al. [40]
34	15	4	3	12	Silberstein et al. [55]
64	54	3	–	7	Riccabona et al. [52]
	547	80	50	97	
	87%	13%	34%	66%	

It is well established that ^{67}Ga is also accumulated in intracranial neoplasms [6, 9, 21, 22, 61] and comparative studies have demonstrated ^{67}Ga uptake in brain tumours which could not be visualized or could hardly be visualized by ^{99}Tc^m scintigraphy, e.g. as shown by Edeling et al. [6] who found that a positive scan could be obtained only by means of ^{67}Ga in 5 out of 11 brain tumours. In cerebral infarcts, however, ^{67}Ga scans were regularly normal despite a clearly delineated accumulation of ^{99}Tc^m. On the basis of these results, ^{67}Ga scanninng is considered a useful adjunct, particularly in cases of tumours that are equivocally detectable or non-detectable with ^{99}Tc^m, as well as in the differentiation of tumour from infarct.

As with strontium nuclides [42], ^{67}Ga is capable of delineating bone neoplasms, especially skeletal metastases, and permits an early diagnosis of bone lesions in the absence of radiological signs of osseous involvement [49]. But with the present state of knowledge, ^{99}Tc^m-labelled phosphate compounds should be preferred for bone scintigraphy.

3. Determinant factors affecting gallium-67 accumulation in tumours

A series of bronchial carcinomas analysed according to histological classifications showed no dependence of ^{67}Ga accumulation on the morphological type of tumour [38, 41]. These clinical results are confirmed by quantitative in-vitro studies of the ^{67}Ga concentration in human tissue samples, summarized in Table II (Kempken et al. [37]). Recently, however, Johnston et al. [33] suggested an apparent relationship between ^{67}Ga affinity and the histopathology as 71% of proven but untreated tumours of nodular sclerosis were detected by means of ^{67}Ga compared with 53% of the mixed cell type. In addition, the high uptake of ^{67}Ga in viable tumour could be demonstrated histologically and autoradiographically, whereas necrotic and fibrotic tumours showed much less ^{67}Ga localization [8, 19]. Corresponding to the fundamental observation of Edwards and Hayes [8], clinical studies (Table III) showed a significantly reduced number of positive scans in patients treated either by chemotherapy or by radiotherapy and a high percentage of positive results (87%) in untreated cases. This is contrary to the results of ORII [48] which showed that ^{67}Ga uptake is independent of tumour growth. During and after ^{60}Co radiotherapy, Higasi and Nakayama [25] showed that the uptake of ^{67}Ga-citrate decreases in proportion to radiation dose. Similarly, Goepfert and Trapp [12] reported a decreasing ^{67}Ga accumulation in tumour tissue during chemotherapy. On the basis of this special feature a valuable test was presented for assessing the activity of the disease. A negative scan does not exclude the presence of viable tumour masses, but a positive result after therapy is a clear proof of either the recurrence of a tumour or the inadequacy of treatment.

CONCLUSION

The use of ^{67}Ga as a tumour-seeking agent for scanning is limited by its lack of tumour-specificity. In summary, the clinical value of ^{67}Ga in tumour scintigraphy is based on the possibility of:

Establishing the extent of a primary tumour, especially one situated in the chest;
Detecting additional tumour foci, particularly metastases previously unknown;
Applying ^{67}Ga for the staging of Hodgkin's disease and other malignant lymphomas, as an important aspect in therapy planning;
Helping in diagnostic differentiation between cerebral vascular lesions and brain tumours;
Indicating residual tumour or recurrence of tumour after therapy;
and
Evaluating the effectiveness of a treatment and susceptibility of tumour to irradiation.

REFERENCES

[1] ANGHILERI, L.J., Studies on the accumulation mechanisms of ^{67}Ga in tumors: role of calcium metabolism, Strahlentherapie 144 (1972) 29.
[2] BERELOWITZ, M., BLAKE, K.C.H., ^{67}Gallium in the detection and localization of tumors, South African Med. J. 45 (1971) 1351.
[3] CELLERINO, A., et al., Operative and pathologic survey of 50 cases of peripheral lung tumors scanned with ^{67}gallium, Chest 64 (1973) 700.
[4] DeLAND, F.H., et al., ^{67}Ga-citrate imaging in untreated primary cancer: preliminary report of cooperative group, J. Nucl. Med. 15 (1974) 408.
[5] DVORAK, K., MORITZ, G., "Possibilités et limites du gallium-67 dans le diagnostic des tumeurs", Medical Radioisotope Scintigraphy 1972 (Proc. Symp. Monte Carlo, 1972) 2, IAEA, Vienna (1973) 681.
[6] EDELING, C.-J., et al., ^{67}Ga-scintigraphy of the brain – a second choice? Nucl. Med. 13 (1974) 144.
[7] EDWARDS, C.L., HAYES, R.L., Tumor scanning with ^{67}Ga citrate, J. Nucl. Med. 10 (1969) 103.
[8] EDWARDS, C.L., HAYES, R.L., Scanning malignant neoplasms with gallium-67, J. Am. Med. Assoc. 212 (1970) 1182.
[9] FOGH, J., EDELING, C.J., ^{67}Ga-scintigraphy of malignant tumors, Nucl. Med. 11 (1972) 371.
[10] FRÖHLICH, G., et al., Zur Bedeutung der Anwendung von ^{67}Ga-Zitrat in der Tumordiagnostik des Thorax, Fortschr. Röntgenstr. 119 (1973) 578.
[11] GLAUBITT, D., et al., Kinetic studies in rats for the determination of the radiation dose by ^{67}Ga, Proc. 2nd Congr. Europ. Ass. Radiol. Amsterdam, 14-18 June 1971, Excerpta Medica.
[12] GOEPFERT, H., TRAPP, P., Abnahme der ^{67}Ga-Zitrat-Einlagerung unter Chemotherapie, Fortschr. Röntgenstr.116 (1972) 126.
[13] GREBE, S.F., et al., Die Gallium 67-Szintigraphie bei der Lungentuberkulose und bei der Silikose, Langenbecks Arch. Chir. Suppl. Forum (1973) 133.
[14] GREBE, S.F., Szintigraphische Diagnostik des Bronchialkarzinoms, Thoraxchir. 19 (1971) 263.
[15] GREBE, S.F., et al., Die Möglichkeiten des Nachweises und der Lokalisation von malignen Tumoren mit der ^{67}Ga-Szintigraphie, Fortschr. Röntgenstr. 116 (1972) 73.
[16] HAMAMOTO, K., et al., Usefulness of computer scintigraphy for detecting liver tumor with ^{67}Ga-citrate and the scintillation camera, J. Nucl. Med. 13 (1972) 667.
[17] HAUBOLD, U., AULBERT, E., "Gallium-67 as a tumour-scanning agent: clinical and physiological aspects", Medical Radioisotope Scintigraphy 1972 (Proc. Symp. Monte Carlo, 1972) 2, IAEA, Vienna (1973) 553.
[18] HAYES, R.L., et al., Factors affecting the localization of ^{67}Ga in animal tumors, J. Nucl. Med. 11 (1970) 324.
[19] HAYES, R.L., EDWARDS, C.L., "New applications of tumor-localizing radiopharmaceuticals", Medical Radioisotope Scintigraphy 1972 (Proc. Symp. Monte Carlo, 1972) 2, IAEA, Vienna (1973) 531.
[20] HEIDENREICH, P., et al., Untersuchungen zur Tumorlokalisation mit ^{67}Ga, Fortschr. Röntgenstr. 115 (1971) 14.
[21] HELLER, H., Hirnszintigraphie mit ^{67}Gallium, Fortschr. Röntgenstr. 117 (1972) 704.
[22] HENKIN, R.E., et al., Adjunctive brain scanning with ^{67}Ga in metastases, Radiology 106 (1973) 595.
[23] HENKIN, R.E., et al., Scanning treated Hodgkin's disease with ^{67}Ga citrate, Radiology 110 (1974) 151.
[24] HIGASI, T., et al., Diagnosis of malignant tumor with ^{67}gallium-citrate (2nd Report), Radioisotopes 19 (1970) 311.

[25] HIGASI, T., NAKAYAMA, J., Clinical evaluation of ^{67}Ga-citrate scanning, J. Nucl. Med. **13** (1972) 196.
[26] HIGASI, T., et al., The mechanism of uptake of gallium-67 in tumor cells, Radioisotopes **22** (1973) 27.
[27] HÖR, G., et al., "Tumorszintigraphie mit ^{67}Ga", 8th Annual Meeting, Society of Nuclear Medicine, Hannover, Sep. 1970 (PABST, H.W., HÖR, G., Eds) Nuklearmedizin, F.K. Schattauer-Verlag, Stuttgart (1972) 318.
[28] HÖR, G., et al., Untersuchungen zum biologischen Verhalten von Radiotechnetium, Radioindium und Radiogallium während der Laktation, Int. J. Appl. Radiat. Isot. **24** (1973) 525.
[29] HÖR, G., et al., Szintigraphische Untersuchungen in der Differentialdiagnose von gutartigen und bösartigen Schilddrüsenerkrankungen mit ^{67}Ga, Acta Endocrinol., Suppl. **179** (1973) 77.
[30] HÖR, G., et al., "Munich report on ^{67}Ga: A review of our experiences in nuclear biology, experimental and clinical nuclear medicine after four years use", Proc. 1st World Congress of Nuclear Medicine, Tokyo and Kyoto (1974).
[31] HUPF, H.B., BEAVER, J.E., Cyclotron production of carrier-free 67-gallium, Int.J.Appl.Radiat.Isot. **21** (1970) 75.
[32] ITO, Y., et al., Diagnostic evaluation of ^{67}Ga-scanning of lung cancer and other diseases, Radiology **101** (1971) 355.
[33] JOHNSTCN, G., et al., ^{67}Ga-citrate imaging in untreated Hodgkin's disease: primary report of cooperative group, J. Nucl. Med. **15** (1974) 399.
[34] KAPLAN, W.D., et al., ^{67}Ga-citrate and the nonfunctioning thyroid nodule, J. Nucl. Med. **15** (1974) 424.
[35] KAY, D.N., McCREADY, V.R., Clinical isotope scanning using ^{67}Ga citrate in the management of Hodgkin's disease, Br. J. Radiol. **45** (1972) 437.
[36] KEMPKEN, K., et al., Nuklearmedizinische Diagnostik von Schilddrüsenmalignomen mit ^{67}Ga, IX. Nuklearmedizin. Symposium Reinhardsbrunn/DDR, Schriftenreihe Wissenschaftliche Tagung in der DDR (1972) 44.
[37] KEMPKEN, K., et al., "Klinisch-experimentelle Untersuchungen zur quantitativen ^{67}Ga-Anreicherung in malignen und nicht malignen Geweben", 11th Annual Meeting, Society of Nuclear Medicine, Athens, Sep. 1973 (PABST, H.W., HÖR, G., Eds), Nuklearmedizin, F.K. Schattauer-Verlag, Stuttgart (1974) 476.
[38] KINOSHITA, F., et al., Scintiscanning of pulmonary diseases with ^{67}Ga-citrate, J..Nucl. Med. **15** (1974) 227.
[39] KRAMER, R.J., et al., Accumulation of gallium-67 in regions of acute myocardial infarction, Am. J. Cardiol. **33** (1974) 861.
[40] LANGHAMMER, H., HÖR, G., HEIDENREICH, P., KEMPKEN, K., PABST, H.W., "Recent advances in tumour scintigraphy using gallium-67", Medical Radioisotope Scintigraphy 1972 (Proc. Symp. Monte Carlo, 1972) **2**, IAEA, Vienna (1973) 607.
[41] LANGHAMMER, H., et al., ^{67}Ga for tumor scanning, J. Nucl. Med. **13** (1972) 25.
[42] LANGHAMMER, H., FREY, K.W., Tumor scintigraphy with radiogallium and radiostrontium, J. Int. Assoc. DMF Radiol. **1** (1972) 29.
[43] LARSON, S.M., et al., The value of ^{67}Ga scanning in the evaluation of liver involvement in Hodgkin's disease: comparison with ^{99}Tc^{m}-sulfur colloid, Nucl. Med. **10** (1971) 241.
[44] LAVENDER, J.P., et al., ^{67}Ga scanning in neoplastic and inflammatory lesions, Br. J. Radiol. **44** (1971) 361.
[45] LOEVINGER, R., BERMAN, M., A schema for absorbed-dose calculations for biologically distributed radionuclides, J. Nucl. Med., Suppl. **1** (1968) Pamphlet 1.
[46] MORI, T., et al., "Clinical results of ^{99}Tc^{m}-labeled bleomycin scintigraphy for tumor imaging", Proc. 1st World Congress of Nuclear Medicine, Tokyo and Kyoto, 1974.
[47] NELSON, B., et al., Distribution of gallium in human tissues after intravenous administration, J. Nucl. Med. **13** (1972) 92.
[48] ORII, H., Tumor scanning with gallium (^{67}Ga) and its mechanism studied in rats, Strahlentherapie **144** (1972) 192.
[49] OKUYAMA, S., et al., Prospects of ^{67}Ga scanning in bone neoplasms, Radiology **107** (1973) 123.
[50] POPHAM, M.G., et al., Evaluation of the dosimetry of intravenously administered ^{67}Ga citrate from measurements of the distribution in male August-Marshall hybrid rats, Br. J. Radiol. **43** (1970) 807.
[51] RAMOS, M., et al., Szintigraphische Tumordiagnostik mit ^{67}Ga-Citrat, Fortschr. Röntgenstr. **117** (1972) 689.
[52] RICCABONA, G., et al., Szintigraphische Erfassung von Malignomen mit ^{67}Ga-Citrat, Nucl. Med. **10** (1971) 234.
[53] SAUNDERS, M.S., et al., The dosimetry of ^{67}Ga citrate in man, Br. J. Radiol. **46** (1973) 456.
[54] SIEMSEN, J.K., et al., Pulmonary concentrations of ^{67}Ga in pneumoconiosis, Am. J. Roentgenol. **120** (1974) 815.
[55] SILBERSTEIN, E.B., et al., ^{67}Ga as a diagnostic agent for the detection of head and neck tumors and lymphoma, Radiology **110** (1974) 605.
[56] SUZUKI, T., et al., Positive scintiphotography of cancer of the liver with ^{67}Ga-citrate, Am. J. Roentgenol. **113** (1971) 92.
[57] SUZUKI, T., et al., Serum alpha-fetoprotein and ^{67}Ga-citrate uptake in hepatoma, Am. J. Roentgenol. **120** (1974) 627.
[58] SWARTZENDRUBER, D.C., et al., Gallium-67 localization in lysomal-like granules of leukemic and nonleukemic murine tissues, J. Natl. Cancer Inst. **46** (1971) 941.
[59] TURNER, D.A., et al., The use of 67-Ga scanning in the staging of Hodgkin's disease, Radiology **104** (1972) 97.

[60] VAIDYA, S.G., et al., Localisation of gallium-67 malignant neoplasms, Lancet 31 (1970) 911.
[61] WAXMAN, A.D., et al., Differential diagnosis of brain lesions by gallium scanning, J. Nucl. Med. 14 (1973) 903.
[62] WINCHELL, H.S., et al., Visualization of tumors in humans using ^{67}Ga-citrate and the Anger whole-body scanner, scintillation camera and tomographic scanner, J. Nucl. Med. 11 (1970) 459.

DISCUSSION

W.H. BEIERWALTES: Have you had enough time or enough data to allow a conjecture on a possible relationship between decreased ^{67}Ga uptake and the response to radiation and/or chemotherapy treatment, for instance a correlation with death or survival rate? If, after treatment, you no longer can get gallium into the tumour, does that mean that you have markedly decreased the death rate from that disease?

H. LANGHAMMER: Yes.

V.R. McCREADY: We have until now studied six patients who had injections of ^{67}Ga once a week over a period of 4 weeks during treatment. We found that the gallium concentration rises steadily in the lesion till about 2 or 3 weeks and in cases of radiotherapy it suddenly drops at about 4 weeks, whereas with chemotherapy it drops at about 3 weeks. It may be that the therapy hampers the gallium uptake into the lesion, but the explanation may also be that the lesion stops growing. The number of cases is small, but we didn't find any relation between the survival and the disappearance rate for the gallium uptake.

T. MUNKNER: If the ^{67}Ga uptake in the mammary glands were studied during chemotherapy or radiotherapy we might learn more about the mechanisms which are responsible for the cell uptake.

V.R. McCREADY: Knowing the study published by Fogh[1], we have been interested in studying whether the mammary cells concentrate Ga or whether Ga is excreted by the glands. The results today seem to show that Ga behaves rather like Ca; we don't find an absolute increased uptake in the cells, and the increased concentration is due to an actual passage of Ga through the cells from the circulation into the milk.

R.L. HAYES: We have had similar results in animal experiments. The interesting thing is that others have shown that there is a definite increase in lysosomal enzyme concentrations in mammary glands during the gestation period and also during the lactation period. In our experiments one-day-old offspring retained a considerable amount of the activity which was secreted in the milk of the mother, whereas nine-day-old offspring lost most of the activity that they had obtained from the milk via the faeces.

V.R. McCREADY: One of my colleagues has plotted the uptakes in various organs and has demonstrated rather nicely that you get high Ga concentrations in secretory organs like the parathyroid, the breast, the gut, etc.

R.M. KNISELEY: It is very difficult to interpret material collected from different clinics. The administered ^{67}Ga dose differs, the same is true about the preparations, contaminants, intervals, instruments, sensitivity, varying detection rate, etc. The results in the tables were indeed boiled down to plus and minus, but the plus does not mean that you have really demonstrated all the possible tumour sites. In addition, the histological label for the tumour may vary from one institute to another, and the criteria for the studies may also vary. You will find a great inter-observer variation in the evaluation or interpretation of the scans, and you certainly have to appreciate the range of normal. Early studies with radiogallium may be interpreted differently today when they are evaluated in retrospect. Finally, you shouldn't forget that the enthusiastic physician goes into detailed studies to show a tumour, if he knows there should be one.

V.R. McCREADY: This is no new problem. We must find a yardstick so that future studies are somewhat more comparable than they are at the moment.

---

[1] Proc. Soc. Exp. Biol. Med. 138 (1971) 1086.

W.H. BEIERWALTES: To obtain a fair evaluation of the utility of Ga uptake and Ga disappearance for choosing the right therapy and for diagnosis, one should establish some kind of joint committee to set up the criteria that should be used all over in such a study. It would be necessary to have reasonable financial support and an adequate number of institutions involved. The study should be concluded within a reasonable period, for instance five years, and the conclusions should be conveyed to all physicians interested in this area.

Good results were obtained by a similar approach in the first systematic study of radionuclide therapy for thyrotoxicosis, conducted by the International Centre for Radiological Health. The financial support was adequate, there was a large group of interested investigators, the criteria for judging the state of disease were laid down, and a number of the biochemical studies were conducted collectively in the same laboratories. It is also reasonable to mention the present US joint committee for defining uniform criteria for the staging of cancer. The job done by this committee has inter alia made it possible to collect all data on thyroid cancer patients on less than one normal disc.

V.R. McCREADY: It's reasonable to stress two observations mentioned in Dr. Langhammer's paper. It is indeed disappointing that there was little relation between the cell type of the cancer and the Ga uptake. We have found in some cases that the more anaplastic the tumour, the better the uptake. For the moment we are just in the middle of a series of seminomas and teratomas, and we find that seminomas will always concentrate gallium, the teratomas don't. The other point of interest is that necrotic tissue doesn't concentrate Ga, but is supposed to concentrate another agent, i.e. Hg.

R.L. HAYES: In our institution, Dr. C.L. Edwards has followed quite a number of patients for a period of years. The patients have routinely been followed at approximately 6-month intervals by repeat ^{67}Ga scans. Although the initial therapy was effective, symptoms reoccurred and very frequently another ^{67}Ga scan was made and definite positive localization was observed. One of these patients is on his fourth re-treatment course now as far as I recall. Dr. Edwards feels that it is most effective and quite rewarding to the patient in terms of his longevity to have repeat ^{67}Ga scans done regularly.

V.R. McCREADY: We, too, never neglect a positive ^{67}Ga uptake. But we don't know the mechanisms behind the uptake. This is one area, I think, which should really be studied by a number of groups.

R.L. HAYES: Let me summarize our current work on the distribution of radiogallium in subcellular fractions, as studied in animal tumours.

About 50% of the total gallium activity is found in both normal and malignant tissue in two fractions, that is in normal lysosomes and in a smaller "lysosome". The smaller particles are present in a very low concentration which has made it extremely difficult to get a pure preparation of these particles because of the many ER vesicles that necessarily contaminate this preparation. Gel filtration experiments indicate that there may be a special macromolecular binding of gallium to a macromolecular species having an approximate molecular weight of 40 000 in aqueous extracts of the tissues. This fraction is also present in the thymus.

The other 50% seems to be divided between the supernatant, a very small amount in mitochondria, and a small amount associated with the nuclei. At the macromolecular level $10 - 15\%$ of the gallium activity is linked with a compound having a molecular weight of about 200 000, whereas $30 - 40\%$ seems to be eluted in the bed volume, which may represent ^{67}Ga activity that has broken free from molecular binding components in the process of separation or gallium associated with a very low molecular weight fraction.

W.H. BEIERWALTES: Does the uptake of gallium depend on the concentration of the macromolecular fraction around 40 000 in the thymus?

R.L. HAYES: We haven't measured the actual amount of the 40 000 molecular weight macromolecular component, we are simply looking at its association with gallium. But we feel very confident at this point that the absolute amount of this 40 000 molecular weight material that binds gallium is very minute. Incidentally, you will get the same 40 000 component when you extract the separated particles, that is the lysosomes and the small-particle fractions. The decrease which I mentioned was referring to the fact that thymus in a normal animal will have about the same percentage of ^{67}Ga associated with the 40 000 molecular weight fraction as is associated with this fraction in tumour tissues. If you implant a tumour in a host animal, sacrifice the animal and process the thymus from it then the amount of the 40 000 molecular weight material associated with Ga is reduced by a factor of about 2 in the thymus.

T. MUNKNER: Do you have any experience with thymus-nude mice?

R.L. HAYES: This special strain is available at the Oak Ridge National Laboratory, but until now we haven't made any experiments on these mice.

V.R. McCREADY: From our studies it seems as if Ga uptake takes place just after DNA synthesis. Gallium uptake may reflect cell division but in a tumour this may not necessarily be division of the tumour cells, it may be the vascular cells or cells of the supportive tissue.

In one of your papers, Dr. Hayes, you have demonstrated that of all the radiopharmaceuticals you have tried, Ga was the best and had the highest uptake in a whole range of animal tumours. Have you any theory explaining why Ga is the best choice?

R.L. HAYES: I have none, except to say that it doesn't appear to be based on growth rate. One of the hepatomas that gives us the highest uptake of Ga is very slow growing, it takes about one month for the tumour to grow to an appreciable size.

TUMOUR LOCALIZATION WITH TECHNETIUM-99m

H.J. GLENN, T.P. HAYNIE, T. KONIKOWSKI
Department of Medicine,
The University of Texas System Cancer Center,
M.D. Anderson Hospital and Tumor Institute,
Houston, Texas,
United States of America

Abstract

TUMOUR LOCALIZATION WITH TECHNETIUM-99m.
A variety of compounds labelled with $^{99}Tc^m$ have been used in procedures whose aim is the localization of tumours because of the availability, the convenience, the cost, the excellent physical characteristics, and the low patient radiation dose of the radionuclide. The most successful of these efforts in the localization of tumours by direct tumour uptake has been in the brain, thyroid and bone. In these, the mechanisms of uptake, simple diffusion, active transport, chemisorption, and/or exchange diffusion, permit rapid accumulation of the tracer by the tumour. The short physical half life of six hours of $^{99}Tc^m$ limits its usefulness when the tumour accumulates the radioactive substance by more protracted mechanisms dependent on prolonged blood concentration and slow renal clearance. $^{99}Tc^m$-labelled compounds have also been used in tumour localization efforts based on the general principles of radionuclide exclusion from the tumour and on altered physiology.

I. INTRODUCTION

There are a variety of $^{99}Tc^m$ compounds that have been used in the localization of tumours. However, these compounds are usually not prominently mentioned in general reviews of tumour-localizing radionuclides [1–4] because these reviews deal with compounds directly taken up by the tumour and ignore tumour localization by exclusion principles, or tumours located by means of altered physiology. The primary clinical use of technetium-labelled compounds for direct tumour localization has been largely confined to brain and thyroid tumours. However, the recent introduction of $^{99}Tc^m$-polyphosphates and $^{99}Tc^m$-diphosphonate compounds has greatly increased the use of $^{99}Tc^m$ for imaging primary bone tumours which apparently localize the tracers. Metastatic lesions to the bone are probably detected by uptake in the reactive osteoid of the involved bone.

II. GENERAL FACTORS RELATING TO TECHNETIUM-99m

Technetium-99m is now the most widely used radionuclide in the practice of nuclear medicine. There are good reasons for this. Among these can be listed the following:

(1) Availability: The ^{99}Mo-$^{99}Tc^m$ nuclide generator is an excellent source of $^{99}Tc^m$, half-life 6 hours. The half-life of ^{99}Mo is 67 hours. The efficient column generator with alumina packing requires the use of enriched ^{98}Mo or "fission moly". However, the methylethyl ketone (MEK) liquid-liquid extraction system makes possible the use of a natural unenriched molybdenum in the neutron activation process.

(2) Physical characteristics: $^{99}Tc^m$ has excellent imaging characteristics for both gamma cameras and for rectilinear scanners. There are excellent gamma photon yields of 140-keV energy and no beta emissions. The radionuclide decays by internal transition.

TABLE I. ESTIMATED ABSORBED RADIATION DOSES FROM TECHNETIUM-99m COMPOUNDS (rad/mCi) [7]

Compound	Total body	Blood	Brain	Liver	Kidneys	Gonads	Marrow	Other
-pertechnetate	0.01–0.02		0.006	0.03–0.07	0.01	0.01–0.04		Stomach 0.04–0.2 Thyroid 0.5 0.1 (blocked)
-sulphur-colloid	0.01–0.02	0.02		0.2–0.4		0.01–0.02	0.02–0.03	Spleen 0.2–0.5
-DTPA	0.2	0.03			0.04	0.01–0.02		Bladder 0.4–0.6
[33] -Sn	0.006				0.036			
[33] -Fe-ascorbic acid	0.009				0.243			
[39] -polyphosphates and -diphosphonate	0.01						0.01	Skeletal 0.045 Bladder, no void 0.49 Bladder, void after 1 h 0.07

(3) Convenience: $^{99}Tc^m$-pertechnetate is eluted in sterile, pyrogen-free saline; additional products are easily prepared because of simple chemistries and laboratory procedures [5]. Synthetic kits (commercial or home-made) add to the simplicity of preparing safe and effective labelled compounds.

(4) Cost [6]: $^{99}Tc^m$-labelled compounds are now among the least expensive of radiopharmaceuticals and have a very high value-to-cost ratio.

(5) Radiation dose: The radiation dose to the patient from $^{99}Tc^m$-compounds is low [7] despite the statistically significant large number of photons available. Table I shows estimated radiation doses to the whole body and critical organs for various $^{99}Tc^m$-labelled compounds.

III. $^{99}Tc^m$-LABELLED COMPOUNDS USED IN TUMOUR LOCALIZATION BY TUMOUR UPTAKE

Because of the general intent of this meeting, this paper is limited to compounds involved in direct tumour uptake. These are shown in Table II.

1. $^{99}Tc^m$-pertechnetate

The principle use of this compound is in the localization of brain tumours [10–14]. A clinical dose of 10–20 mCi is used. The mechanism of action is membrane permeability, or simple diffusion [11]. Brain scanning is a widely used neurological and presurgical diagnostic and screening procedure. Usually, uptake in the choroid plexus is minimized by the oral administration of perchlorate. The imaging procedure has a clinical reliability of about 80–90% with false negatives appearing about 10% of the time. In this respect, it is comparable to the use of ^{203}Hg-chlormerodrin or other brain-scanning agents, but is preferred because of its high photon density and low radiation dose to the patient.

$^{99}Tc^m$-pertechnetate has been reported to localize in thyroid tumours, in situ in the thryoid gland and functioning metastases [14–17]. In some cases, the uptake of $^{99}Tc^m$ has not been accompanied by similar findings on ^{131}I scanning, suggesting that the tumours trap, but do not organify, these tracers [18–20]. In other cases, metastases visualized as active areas of uptake on delayed (24 hours or longer) ^{131}I scans have not been visualized on early $^{99}Tc^m$ scans (less than 1 hour after injection), suggesting that high body background may have obscured the picture. The usual dose administered is 1–2 mCi and it's mechanism of localization is that of active transport [13]. It should, however, be pointed out that the vast majority of thyroid neoplasms localize both $^{99}Tc^m$ and ^{131}I less well than does normal thyroid tissue, and are therefore seen as "cold areas" on thyroid scans.

$^{99}Tc^m$-pertechnetate has also been used in the localization of tumours in the breast [20–21]. The usual dose administered is around 10 mCi. Its mechanism of localization is related to increased tumour vascularity and capillary permeability, diffusion. The clinical reliability of the scanning procedure is not high and usually results in about 20% false positive scans. Its value as a screening or confirmatory test has been questioned, but it has been described as being useful as a supplemental approach or in resolving problem cases unanswered by mammography or other diagnostic procedures for breast cancer.

Labelled pertechnetate has also been used in the localization of Warthin's tumour in the salivary glands [22–24]. Although doses from 0.5–15 mCi have been used, the usual dose is 3–10 mCi. The mechanism of tumour localization is that of active transport [11]. The salivary duct cell component of Warthin's tumour accumulates the pertechnetate. The tumour concentrates about five times more radioactivity than does the normal parotid gland. This imaging procedure is not of exceptional value from a localization standpoint as seldom can tumours be

TABLE II. TECHNETIUM-99m COMPOUNDS USEFUL IN DIRECT TUMOUR LOCALIZATION

Compound	Tumour	Dose, i.v. (mCi)	Mechanization of localization	Comments
-TcO$_4^-$	Brain	10–20	Diffusion	Universally used, ClO$_4^-$ Clinical reliability, 80–90%
	Thyroid	1–2	Active transport	Adjunct to ^{131}I
	Breast	10	Diffusion	Clinical reliability poor Adjunct to mammography
	Salivary gland (Warthin's tumour)	3–10	Active transport	Tumour is rare and usually clinically evident
	Eye	5–10	Diffusion?	Reliability (?)
	Spine	15	Diffusion?	Reliability (?)
	Extra-cranial	3–5	Varied	Reliability limited
-Fe-ascorbic acid	Brain	10–20	Diffusion	No perchlorate predose
-Fe-ascorbic acid-DTPA	Brain	10–20	Diffusion	Same reliability as -TcO$_4^-$
-Sn-DTPA	Brain	10–20	Diffusion	Faster renal clearance
-pyrophosphate	Bone	10	Chemisorption	More reliability than X-ray
-polyphosphate	Bone	10	Chemisorption	Stability problem
-diphosphonate	Bone	10	Chemisorption and exchange diffusion	Delineate osteosarcoma well
	Extra-osseous Breast Lung Brain	10	Various (?)	Reliability in question

detected that are not clinically evident. The procedure does help in localization before surgery. Other tumours of the salivary gland may be located by pertechnetate exclusion, i.e. they appear as areas of decreased uptake against the background of normal parotid uptake of pertechnetate.

An attempt to locate orbital tumours of the eye, using $^{99}Tc^m$-pertechnetate, has also been reported [25]. A dose of 0.13 mCi/kg body weight was used. The mechanism of action is not known, but is probably simple diffusion. Not enough work has been done to establish clinical reliability, but in one study, pertechnetate was administered to 21 persons, seven of whom had orbital tumours as proved by photography, and one was questionable. There was increased tumour uptake over the eye tumour in six of these with one questionable. Those subjects without ocular tumours gave balanced scintigrams over each eye.

Pertechnetate has also been used in the location of spinal tumours, using a dose of about 15 mCi [26]. Again, the mechanism of action is not known, but it may be simple diffusion. Body background was blocked by placing strips of lead along each side of the spinal cord. Fifteen patients with suspected spinal tumours showed an increased uptake in the area of the suspected tumours. Areas of increased uptake were not noted in spinal scans of normal individuals. Again, not sufficient work has been done to establish clinical reliability.

Extra-cranial neoplasms have been noted in patients undergoing routine brain scanning with $^{99}Tc^m$-pertechnetate [27]. As a result of this, additional photoscanning with $^{99}Tc^m$ has been undertaken in a series of cases with known extra-cranial tumours, using 3–5 mCi of activity. In the best documented study, 17 of 26 known neoplasms scanned were demonstrated by increased uptake of activity relative to the tumour bed and the surrounding tissues. However, the method has clinical limitations because of the presence of marked activity in normal blood pools and some normal organs, as well as in certain non-neoplastic lesions.

2. $^{99}Tc^m$-iron-ascorbic acid, with and without DTPA

These compounds have been listed together because of the similarity of their biological action and distribution. They are used in brain tumour localization studies [28–29]. They are administered in 10–20 mCi doses, and the mechanism of tumour concentration is that of simple diffusion [11]. Clinical reliability for these agents in brain scanning is about the same as that associated with pertechnetate, even though animal studies using a mouse brain tumour system [28] indicate greater tumour uptake with these substances than with pertechnetate. The principal clinical advantage is that perchlorate does not have to be administered to block uptake by the choroid plexus.

3. $^{99}Tc^m$-Sn-DTPA

This compound is listed independently of the previous two because of its different biological characteristics [31–33]. In the mouse brain tumour system previously mentioned [30], this substance is taken up by the tumour to a less extent than the two previously discussed compounds. It is also cleared by the kidney at a considerably faster rate. Its principal use in tumour localization, again, is associated with tumours of the brain. It has been used in 10–20 mCi doses, and accumulates in the tumour by simple diffusion. Its clinical reliability seems to be about that of other brain scanning agents, and one does not have to block the choroid plexus with perchlorate.

4. $^{99}Tc^m$-pyrophosphate, $^{99}Tc^m$-polyphosphate, and $^{99}Tc^m$-diphosphonates

These compounds are now the most widely used compounds for localizing tumours in bone [34–37]. There is apparently direct uptake of the tracers by small primary bone tumours. Metastatic lesions to the bone are probably detected by uptake in the reactive osteoid of the involved bone. The administered dose of these compounds is about 10 mCi. The mechanisms of action of all are related to bone vascularity (blood flow). The pyrophosphates and polyphosphates probably localize by chemisorption and exchange diffusion, while the diphosphonate may localize more by chemisorption [11]. Growing bone seems to be a prerequisite to tracer accumulation. The clinical reliability of these bone-scanning agents is considered to be good. They detect questionable areas earlier than diagnostic X-ray [38]. Osteosarcomas are delineated with considerable clarity [39].

These compounds have also shown tumour localization outside the skeletal system [40–41]. Soft tissue calcific deposits also accumulate tracer. Extra-osseous metastases from osteosarcoma have also been shown to take up polyphosphates, and diphosphonates have been shown to localize primary breast carcinoma [42–44]. There are differences in opinion as to the value of the diphosphonates in delineating breast carcinoma. These compounds have also been shown to accumulate occasionally in brain tumours [45]. It has also been proposed that focal uptake of polyphosphates and diphosphonates in bone and soft tissues may be due to their binding to receptor sites on enzymes such as acid phosphatase [46].

5. $^{99}Tc^m$-labelled antibiotics

It perhaps should be mentioned that some of the more recent attempts to prepare tumour-localizing compounds labelled with $^{99}Tc^m$ have been in the field of antibiotics, namely $^{99}Tc^m$-tetracycline and $^{99}Tc^m$-bleomycin. The use of these compounds is described in another paper in these Proceedings.

IV. TUMOUR LOCALIZATION BY PROCESSES OTHER THAN DIRECT TUMOUR UPTAKE

Although this is not the primary interest of this meeting, it should be noted that the use of compounds labelled with $^{99}Tc^m$ has been helpful in localizing neoplasms which do not take up the radioactive material. The radioactive material is taken up by the normal tissue, and the neoplasm appears as an area of decreased activity within the organ. Prime examples of this method of tumour localizations are the use of pertechnetate in the scanning for thyroid and gastric malignancies, the use of labelled colloids in liver-spleen scanning, and the use of labelled DTPA compounds in kidney scanning. Because tumours are seen as cold areas, this severely limits the minimum size of the tumour that can be detected. The most successful of these procedures has been the detection of primary and metastatic tumour in the liver.

$^{99}Tc^m$-compounds have also aided in the localization of tumours by means of altered physiology. The most important examples of this are the use of $^{99}Tc^m$-macroaggregated albumin, $^{99}Tc^m$-microspheres, or $^{99}Tc^m$-iron hydroxide in lung scanning or the use of $^{99}Tc^m$-sulphur colloid in lymphatic scanning. Occasionally, more than one general process is involved.

V. SUMMARY

The most successful use of $^{99}Tc^m$-labelled compounds in the localization of tumours by direct uptake has been in the fields of brain, thyroid and bone. In these, the mechanism of uptake, simple diffusion and chemisorption, and/or exchange diffusion, permits rapid uptake of the tracer. The short physical half-life of six hours of $^{99}Tc^m$ limits its usefulness in tumour localization by more protracted mechanisms which seem to depend on blood concentration and slow renal clearance. $^{99}Tc^m$-labelled compounds have also been used in tumour localization efforts which are based on radionuclide exclusion from the tumour or on the detection of altered physiology.

REFERENCES

[1] McCREADY, V.R., TAYLOR, D.M., TROTT, N.G., CAMERON, C.B., FIELD, E.O., FRENCH, R.J., PARKER, R.P., editors, Radioactive Isotopes in the Localization of Tumors (Proc. Int. Nucl. Med. Symp. London, 1967) William Heinemann Medical Books, Ltd. (1969).
[2] HOFFER, P.B., GOTTSCHALK, A., Tumor scanning agents, Sem. Nucl. Med. 4 3 (1974) 305.
[3] BENUA, R.S., Tumor localization with radionuclides, Clin. Bulletin (1973) 49.
[4] HAYES, R.L., EDWARDS, C.L., New Applications of Tumor-Localizing Radiopharmaceuticals, Medical Radioisotope Scintigraphy 1972 Vol.II, IAEA, Vienna (1973) 531.
[5] LATHROP, K.A., Preparation and control of ^{99m}Tc-radiopharmaceuticals, Radiopharmaceuticals From Generator-Produced Radionuclides, IAEA, Vienna (1971) 39.
[6] TOUYA, J.J., BONOMI, R., FERRANDO, R., TOUYA, E., Practical problems in the selection of generators and radiopharmaceuticals, Radiopharmaceuticals From Generator-Produced Radionuclides, IAEA, Vienna (1971) 15.
[7] HINE, G.J., JOHNSON, R.E., Absorbed dose from radionuclides, J. Nucl. Med. 11 7 (1970) 468.
[8] O'MARA, R.E., MOZLEY, J.M., Current status of brain scanning, Sem. Nucl. Med. 1 1 (1971) 7.
[9] WITCOFSKI, R.L., MAYNARD, C.D., ROPER, T.J., A comparative analysis of the accuracy of the technetium-99m pertechnetate brain scan: Followup of 1000 patients, J. Nucl. Med. 8 3 (1967) 187.
[10] BROOKS, W.H., MORTARA, R.H., PRESTON, D., The clinical limitations of brain scanning in metastatic disease, J. Nucl. Med. 15 7 (1974) 620.
[11] KRISHINAMURTHY, G.T., MEHTA, A., TOMIYASU, U., BLAHD, W.H., Clinical value and limitations of ^{99m}Tc brain scan: An autopsy correlation, J. Nucl. Med. 13 6 (1972) 373.
[12] QUINN, J.L., CIRIC, I., HAUSER, W.N., Analysis of 96 abnormal brain scans using technetium-99m (pertechnetate form), JAMA 194 (1965) 157.
[13] SOLOWAY, A.H., DAVIS, M.A., Survey of radiopharmaceuticals and their current status, J. Pharm. Sciences 63 5 (1974) 647.
[14] STEINBERG, M. CAVALIERI, R.R., CHOY, S.H., Uptake of technetium 99-pertechnetate in a primary thyroid carcinoma: Need for caution in evaluating nodules, J. Clin. Endocr. 31 (1970) 81.
[15] WIENER, S.N., GHOSE, M.K., Thyroid metastases diagnosed by ^{99m}Tc scanning, J. De L'Association Canadienne Des Radiologistes, 21 (1970) 190.
[16] SPARAGANA, M., LITTLE, A., KAPLAN, E., Rapid evaluation of thyroid nodules using ^{99m}Tc-pertechnetate scanning, Letter to the Editor, J. Nucl. Med. 11 5 (1970) 224.
[17] DOS REMEDIOS, L.V., WEBER, P.M., JASKO, I.A., Thyroid scintiphotography in 1,000 patients: Rational use of ^{99m}Tc and ^{131}I compounds, J. Nucl. Med. 12 10 (1971) 673.
[18] USHER, M.S., ARZOUMANIAN, A.Y., Thyroid nodule scans made with pertechnetate and iodine may give inconsistent results, J. Nucl. Med. 12 3 (1971) 136.
[19] MEIGHAN, J.W., DWORKIN, H.J., Failure to detect ^{131}I positive thyroid metastases with ^{99m}Tc, J. Nucl. Med. 11 4 (1970) 173.
[20] CANCROFT, E.T., GOLDSMITH, S.J., CONKLIN, E., ^{99m}Tc-pertechnetate scintigraphy as an aid in the diagnosis of breast masses, Radiology 106 (1973) 441.
[21] VILLARREAL, R.L., PARKEY, R.W., BONTE, F.J., Experimental pertechnetate mammography, Radiology 111 (1974) 657.

[22] STEBNER, F.C., EYLER, W.R., DuSAULT, L.A., BLOCK, M.A., Identification of Warthin's tumors by scanning of salivary glands, Am. J. Surgery 116 (1968) 513.
[23] SCHALL, G.L., DI CHIRO, G., Clinical usefulness of salivary gland scanning, Sem. Nucl. Med. 2 3 (1972) 270.
[24] BARON, J.M., ROSEN, G., "Study of the salivary glands by sequential scanning: the radiosialogram", Medical Radioisotope Scientigraphy 1972 (Proc. Symp. Monte Carlo, 1972) 2, IAEA, Vienna (1973) 141.
[25] HEVER, H.E., EHLERS, N., Orbitography with technetium-99m in diagnosis of orbital tumors, Ann Ocul. 205 (1972) 283.
[26] FAZIO, C., AGNOLI, A., BAVA, G.L., BOZZAO, L., FIESCHI, C., Demonstration of spinal tumors with intravenously injected pertechnetate-99mTc: A new diagnostic technique, J. Nucl. Med. 10 (1969) 508.
[27] WHITLEY, J.E., WITCOFSKI, R.L., BOLLIGER, T.T., MAYNARD, C.D., Tc ^{99m} in the visualization of neoplasms outside the brain, The Am. J. of Roentgenology, Radium Ther. & Nucl. Med. XCVI 3 (1966) 706.
[28] STAPLETON, J.E., ODELL, R.W., Technetium/iron/ascorbic acid complex; a good brain scanning agent, Am. J. Roentgenology, Radium Ther. & Nucl. Med. CI 3 (1967) 152.
[29] BROOKEMAN, V.A., WILLIAMS, C.M., Evaluation of ^{99m}Tc-DTPA acid as a brain scanning agent, J. Nucl. Med. 11 (1970) 733.
[30] KONIKOWSKI, T., GLENN, H.J., HAYNIE, T.P., Renal clearance and brain tumor localization in mice of ^{99m}Tc compounds of (Sn) DTPA, (iron-ascorbic acid) DTPA, and iron-ascorbic acid, J. Nucl. Med. 13 1 (1972) 834.
[31] ECKELMAN, W.C., RICHARDS, P., Instant ^{99m}Tc-DTPA, J. Nucl. Med. 11 (1970) 761.
[32] ATKINS, H.L., CARDINALE, K.G., ECKELMAN, W.C., et al., Evaluation of 99mTc-DTPA prepared by three different methods, Radiology 98 (1971) 674.
[33] ECKELMAN, W.C., RICHARDS, P., HAUSER, W., et al., ^{99m}Tc-DTPA preparations, J. Nucl. Med. 12 (1971) 699.
[34] BAKSHI, S., ACKERHALT, R.E., BLAU, M., PARTHASARATHY, K.L., BENDER, M., Routine Bone Scanning for Occult Metastases, Department of Nuclear Medicine, Roswell Park Memorial Institute, New York State Department of Health, Buffalo, New York (1973).
[35] SUBRAMANIAN, G., McAFEE, J.G., BELL, E.G., BLAIR, R.J., O'MARA, R.E., RALSTON, P.H., 99mTc-labeled polyphosphate as a skeletal imaging agent, Radiology 102 (1972) 701.
[36] SUBRAMANIAN, G., McAFEE, J.G., BLAIR, R.J., MEHTER, A., CONNOR, T., 99mTc-EHDP: A potential radiopharmaceutical for skeletal imaging, J. Nucl. Med. 13 12 (1972) 947.
[37] ECKELMAN, W.C., REBA, R.C., KUBOTA, H., STEVENSON, J.S., 99mTc-pyrophosphate for bone imaging, J. Nucl. Med. 15 4 (1974) 279.
[38] SILBERSTEIN, E.B., SAENGER, E.L., TOFE, A.J., ALEXANDER, G.W., PARK, H-M., Imaging of bone metastases with ^{99m}Tc-Sn-EHDP (diphosphonate), 18F, and skeletal radiography, Radiology 107 (1973) 551.
[39] PENDERGRASS, H.P., POTSAID, M.S., CASTRONOVO, F.P., Clinical use of diphosphonate-99mTc (HEDSPA), Radiology 107 (1973) 557.
[40] GHAED, N., THRALL, J.H., PINSKY, S.M., JOHNSON, M.C., Detection of estraosseous metastases from osteosarcoma with 99mTc-polyphosphate bone scanning, Radiology 112 (1974) 373.
[41] FLOWERS, W.M., 99mTc-polyphosphate uptake within pulmonary and soft-tissue metastases from osteosarcoma, Radiology 112 (1974) 377.
[42] BERG, G.R., KALISHER, L., OSMOND, J.D., et al., 99mTc-diphosphonate concentration in primary breast carcinoma, Radiology 109 (1973) 393.
[43] McDOUGALL, I.R., PISTENMA, D.A., Concentration of 99mTc diphosphonate in breast tissue, Radiology 112 (1974) 655.

[44] SIEGEL, M.E., FRIEDMAN, B.H., WAGNER, JR., H.N., A new approach to breast cancer, breast uptake of ^{99m}Tc-HEDSPA, JAMA 229 (1974) 1769.
[45] JONES, A.E., FRANKEL, R.S., DI CHIRO, G., JOHNSTON, G.S., Brain-scintigraphy with ^{99m}Tc pertechnetate, ^{99m}Tc polyphosphate and ^{67}Ga citrate, Radiology 112 (1974) 123.
[46] SCHMITT, G.H., HOLMES, R.A., ISITMAN, A.T., et al., A proposed mechanism for ^{99m}Tc-labeled polyphosphate and diphosphonate uptake by human breast tissue, Radiology 112 (1974) 733.

DISCUSSION

W.H. BEIERWALTES: The pyrophosphate is split very rapidly by phosphatases in bone, whereas the diphosphonate is not. As a result of this difference pyrophosphate disappears from bone in two to four hours, whereas diphosphonate is rather permanent in its association with bone. It's not clear to me, however, why the phosphatases do not cleave diphosphonate.

H.J. GLENN: It may be that the mechanism of absorption depends on phosphatases already absorbed to bone at different sites, and that the polyphosphates, regardless of their nature, are attracted to the phosphatases. There may be some action after the initial attraction on one compound that is not present with the other.

T. MUNKNER: If half of the members of the panel had been oncologists, the discussion might have been more influenced by comments from biologists who are mainly concerned with the natural history of tumours.

Your studies, Dr. Glenn, have been carried out on mice on a standard diet, i.e. they have received a substantial supply of iodide per day. In this case I wonder what can be achieved by an additional intake of perchlorate. Do you have figures on the thyroid iodide uptake in the mice which didn't receive perchlorate? Do you think that perchlorate has actions which are not mimicked by iodide? And, finally, did you use atropine in any of your groups?

H.J. GLENN: No.

K. HISADA: At the World Federation of the Nuclear Medicine Meeting in Tokyo, October 1974, Dr. Benes from Switzerland claimed that ^{99}Tc^m-citrate was a powerful agent for tumour localization. It was useful for brain scintigraphy and also for tumours of soft tissues in the neck and of the extremities, but not for abdominal tumours. Before we heard about the results from Switzerland we had already studied the behaviour of Tc-citrate, Tc-bleomycin and Tc-pertechnetate in animals. As to per cent dose uptake per gram tissue the three compounds show more or less the same results. The difference lies in the blood clearance where Tc-citrate clears the fastest and pertechnetate the slowest. That means that the ratios of tumour to blood, tumour to muscle, tumour to liver are the best for the Tc-citrate, but correspondingly the tumour to kidney ratio is the worst.

H.J. GLENN: When you made this Tc-citrate, did you use equivalent amounts of tin and citric acid, really starting with a tin-citrate compound?

K. HISADA: We followed two different approaches: in one we used tin as the reducing agent and in the other we used sodium borohydrate. The results were better with tin chloride.

R.L. HAYES: Do you, Dr. Hisada and Dr. Glenn, know how Tc-citrate might conceivably stand in relation to other tumour-localizing agents?

H.J. GLENN: It is always dangerous to extrapolate directly from animal experiments to human studies. From our experiments up till now it doesn't look as if there will be any considerable advantage of the new compound over pertechnetate, but we should like to compare the new compound with all the other fifteen or so brain-scanning agents that we have looked at in a very carefully controlled study.

IAEA-MG-50/11

TUMOUR LOCALIZATION WITH RADIONUCLIDES OF INDIUM

H.J. GLENN, T.P. HAYNIE, T. KONIKOWSKI
Department of Medicine,
The University of Texas System Cancer Center,
M.D. Anderson Hospital and Tumor Institute,
Houston, Texas,
United States of America

Abstract

TUMOUR LOCALIZATION WITH RADIONUCLIDES OF INDIUM.
Indium-113m and indium-111 radionuclides have been used in tumour localization efforts. A successful use of ^{113}In^m-labelled compounds in the localization of tumours by direct uptake has been the use of the -DTPA chelate in brain tumour investigations. It appears to localize by the simple rapid mechanism of diffusion, which is compatible with the short half-life of the radionuclide. The clinical reliability of ^{113}In^m-DTPA in brain tumour diagnosis has been reported to be about the same as that of ^{99}Tc^m pertechnetate. ^{111}In chloride, by direct administration or pre-bound to autologous serum, is the form of ^{111}In most intensively studied in tumour uptake investigations. It has shown excellent uptake in certain animal tumour systems. Its use in general tumour scanning has been limited. In comparative studies, ^{67}Ga-citrate has seemed to be superior as a general tumour-scanning agent. However, the use of ^{111}In-chloride as a brain-scanning agent and for tumours in the head and neck region, in doses of 3 - 5 mCi, seems to warrant further investigation.

I. INTRODUCTION

Because of the close association of indium and gallium in the periodic table, it is not surprising that, based on the success of gallium, there is interest in the use of indium for the localization of tumours. In attempting to determine the status of indium-labelled compounds in tumour localization, it is necessary to consider two radionuclides of indium, ^{113}In^m and ^{111}In. Although mention may be found of other indium radionuclides, their roles have been relatively unimportant. Most studies with ^{113}In^m-labelled compounds have been in humans, while much of the information concerning ^{111}In-labelled compounds is in animals. Because of this, but also because of differences in physical radiation properties, modes of production and other parameters, these two radionuclides are presented independently.

II. GENERAL FACTORS

Radiopharmaceuticals labelled with ^{113}In^m have been widely used. Although ^{99}Tc^m is preferred particularly for use with gamma cameras, ^{113}In^m-labelled compounds are used in many areas particularly where ^{99}Tc^m is not readily available. Let us consider the following factors relating to ^{113}In^m.

1. Availability: The radionuclide ^{113}In^m is secured from the ^{113}Sn-^{113}In^m nuclide generator [1]. The half-life of the parent ^{113}Sn is 115 d; the half-life of the ^{113}In^m is 1.66 h. An efficient generator can be prepared using hydrous zirconium oxide as the absorbing material. It requires the use of enriched ^{112}Sn for the preparation of the radioactive tin by neutron activation.

TABLE I. ESTIMATED ABSORBED RADIATION DOSES FROM INDIUM-113m AND INDIUM-111 COMPOUNDS (rads/mCi)

Compound	Total body	Blood	Liver	Marrow	Bladder	Other
^{113}In^m-DTPA [5]	0.01–0.02	0.03			0.4	
^{113}In^m-DTPA [4]	0.009	0.034			0.5–0.6	Kidneys 0.06
^{113}In^m-EDTA	0.003				0.18	Gonads 0.027
^{111}In-chloride [28]	0.5		4.5	3.6		
^{111}In-transferrin [7]	0.14		2.4	2.7		

2. Physical characteristics: ^{113}Sn decays by electron capture to ^{113}In^m which decays by isomeric transition to give gamma photons with an energy of 393 keV, and no beta emissions. However, there is about 35% internal conversion to conversion electrons. The 393-keV gamma has about the same energetics as the photon peak used for ^{131}I. A high energy collimator is used in the scanning procedure.

3. Convenience: The long half-life of the ^{113}Sn makes possible long use of the generator (6-8 months). This is an appealing characteristic for use in areas where weekly ^{99}Mo-^{99}Tc^m generators are not available or prohibitively expensive. The generator is available in sterile or non-sterile form. The indium is eluted with 0.05N HCl, which is injectable only in very small quantities. Simple chemistry is involved in the making of other labelled compounds [2]. When compared with ^{99}Tc^m, only relatively small generators (5-25 mCi) are available, but the generator may be milked two and perhaps three times in a day to give additional quantities of material.

4. Cost [3]: The initial cost of such generators is high largely because enriched ^{112}Sn must be used in the neutron activation process. However, its long useful half-life of 6-8 months makes the per-patient cost fairly low, with a good value-to-cost ratio.

5. Radiation dose: There is low radiation dose [4-6] to the patient from the use of this radionuclide. The absorbed radiation doses to the whole body and critical organs for compounds used in tumour localization are shown in Table I.

The picture with ^{111}In is quite different from that for ^{113}In^m. The following factors relate to ^{111}In.

1. Availability: Since this radionuclide is cyclotron-produced, it is available only near a cyclotron or where logistics from the cyclotron to the user are favourable. The most common way of preparing the radionuclide is from enriched cadmium using the reaction ^{111}Cd(p,n)^{111}In. It has also been prepared by the alpha bombardment of natural silver using the reaction $^{108-110}$Ag (α,xn) ^{111}In.

2. Physical characteristics: Its physical characteristics are most desirable. It decays more than 99% by electron capture to ^{111}Cd. It has a half-life of 2.82 d, and gives a 173-keV gamma followed in cascade by a 247-keV gamma.

3. Convenience: Some producers of cyclotron radionuclides supply ^{111}In-chloride in 0.05N HCl, chemical grade. Radiopharamaceutical producers may supply both the ^{111}In-chloride and the ^{111}In-DTPA compounds in sterile, pyrogen-free solution. Only small volumes of the chloride in 0.05N HCl are injectable.

4. Cost: The radionuclide is generally considered one of the more expensive radionuclides, the pharmaceutical forms costing approximately US $15.00 per mCi at present prices.

5. Radiation dose: The radiation dose from the nuclide may be high, particularly if the form administered localizes irreversibly in one or more organs or binds to serum proteins, e.g. transferrin. Table I gives the estimated radiation doses from substances labelled with ^{111}In [7].

III. LABELLED INDIUM COMPOUNDS USED IN DIRECT TUMOUR LOCALIZATION

Although indium compounds have also been used in tumour localization by indirect means (exclusion or altered physiology) the discussion in this paper is largely limited to those which show direct accumulation in tumours to a greater extent than in the surrounding tissues.

1. Substances labelled with ^{113}In^m

Because of the short half-life of ^{113}In^m, labelled substances used in the localization of tumours by direct radionuclide uptake must be limited to those with rapid mechanisms of localization. The ^{113}In^m chelates used in brain tumour localization studies meet this requirement. The EDTA or DTPA chelates are easily prepared in the presence of carrier ferric ion by adding EDTA [8] or DTPA [9-10] to an acid solution of the ions and then neutralizing with a base or buffer. O'Mara and associates [4] studied the uptake of ^{99}Tc^m-TcO$_4^-$ and ^{113}In^m-labelled -DTPA, -cysteine, -tryptophane and -nitrate in a mouse ependymoblastoma. The absolute tumour uptake of the ^{113}In^m-DTPA was about the same as that of pertechnetate at 15 min, and about half that of pertechnetate at 1 h. However, at 1 h, the tumour-to-brain ratio of about 30 for the ^{113}In^m-DTPA was about twice the ratio for pertechnetate. The subcellular distribution of the chelate resembled that of pertechnetate.

Haynie and co-workers [11] have studied ^{113}In^m-DTPA and ^{99}Tc^m-pertechnetate in an in-situ mouse brain sarcoma system. The maximum tumour uptake for ^{113}In^m-DTPA was 2.18% dose/g tumour while that of pertechnetate was 3.93% dose/g tumour by itself and 5.14% dose/g tumour with perchlorate predose. The maximum tumour-to-brain ratio for ^{113}In^m-DTPA was 10.0, that for pertechnetate alone was 7.3, and that for pertechnetate with perchlorate predose was 9.7.

There are reports of the use of ^{113}In^m-DTPA in human brain tumour localization studies [4, 12, 13]. It is administered in doses of 3 - 20 mCi, but because of the limited size of the nuclide generator, 3 - 5 mCi doses are more common. It localizes by simple diffusion [14] and does not localize in the choroid plexus, thyroid or salivary glands, as with ^{99}Tc^{m}-pertechnetate. The clinical reliability is about the same as that of ^{99}Tc^m-pertechnetate, but it may be better in the infratentorial regions [15].

Cuaron et al. [16] have reported the demonstration of abnormal uptake of ^{113}In^m-chloride in tumours by delayed hepatic scintiscanning. A 10-min scan gave a picture of the hepatic blood pool. A follow-up scan at 6 h showed abnormal uptake in 72% of the primary tumours and 70% of the metastatic tumours. A 10-mCi dose was employed. They postulate a mechanism whereby circulating iron slowly displaces the ^{113}In^m from the transferrin to which it is bound on injection, and the freed ionic indium then binds to secondary iron binding sites in tumour and other iron-storage areas.

2. Compounds of ^{111}In

^{111}In-citrate has been used in animal studies involving rats and mice. It is usually prepared by adding ^{111}In-chloride to an excess of citrate ion in solution. Wagner et al. [17] studied the uptake of ^{111}In-citrate in three mouse tumours, and reported that the tumour-to-organ ratios were less than one for lung, liver, spleen, kidney and bone at 24 and 48 h. They obtained good scans of mastocytomas implanted in the hind legs of mice, but scans did not demonstrate Sarcoma 180 transplanted in the chest wall of mice. The successful scans were attributed to the large size of the mastocytoma, which was 25% of the total body weight of the animal and accumulated 29% of the injected dose. Additional work with ^{59}Fe and ^{114}In^m in paired animals suggested that nonspecific change was the mechanism for tumour localization. Hayes and Edwards [18] compared ^{111}In-citrate with ^{67}Ga-citrate in three rat and three mouse tumours. Some of the studies were done at 1 d and some at 3 d. ^{111}In-citrate was superior in per cent uptake by tumour and tumour-to-liver, tumour-to-spleen, tumour-to-muscle and tumour-to-blood ratios in Walker 256 carcinoma, but was inferior in its localization in the other five tumours. Merrick et al. [19] have studied the biological distribution and uptake in sterile granulomas of ^{111}In-citrate, -fluoride, -acetate, -lactate, -HEDTA chelates as part of their comparative study with ^{111}In-bleomycin and ^{67}Ga-citrate. They conclude that there is no advantage in using compounds of ^{111}In instead of ^{111}In-chloride.

^{111}In-chloride has been widely studied in animal systems and in man. Because the indium apparently binds almost immediately on injection to serum proteins, largely transferrin, the use of the chloride form is occasionally referred to as ^{111}In-transferrin. This may cause confusion as to the exact compound used; good pharmaceutical and medical practice recommends giving the name of the compound injected regardless of biological affinity, therefore, ^{111}In-chloride should be the preferred designation. In this review, both the labelled chloride and the labelled transferrin are considered as the same compound with special note being made where the labelled protein was actually administered.

Serafini et al. [20] have studied the uptake of ^{111}In-chloride in spontaneous lymphosarcoma R2788 in rats before and after exposure to irradiation therapy. When compared with the uptake of ^{67}Ga-chloride, ^{111}In proved to be the more favourable radionuclide with 3.9% of the administered dose/g accumulating in the tumour at 48 h. A rapid disappearance of the palpable tumour masses following irradiation therapy was accompanied by a decrease in radiopharmaceutical uptake.

Higasi et al. [21] have studied the tumour uptake of three radionuclides simultaneously injected into mice with Ehrlich's tumour transplanted into the femoral region. A semiconductor detector was used in tumour uptake measurements. The per cent dose/g tumour uptake at 48 h of ^{111}In-chloride was about 3.3%; that of ^{67}Ga-citrate was 4.6%. ^{111}In-chloride was therefore judged somewhat inferior to ^{67}Ga-citrate in this study.

Konikowski and co-workers [22, 23] have also studied the uptake of ^{111}In-labelled compounds in their in-situ mouse brain tumour animal model. With the ^{111}In-chloride, they have reported a maximum per cent dose/g tumour uptake of 18.5%, and a maximum tumour-to-brain ratio of 17%, the highest reported for any one of 14 radiopharmaceuticals studied in their system. The overall imaging characteristics of ^{111}In-chloride were generally superior at most time intervals studied. The renal blood clearance of the ^{111}In-chloride in mice was only 0.0022 ml per min, the slowest of any of the radiopharmaceuticals studied. This results in a high total body absorbed radiation dose estimated in the human to of this order of 0.395 rads/mCi injected [24].

Zeidler and co-workers [7] studied the kinetics and distribution of ^{111}In which had been bound to autologous serum before injection in rabbits and human. The labelled transferrin (or the radionuclide released from it) accumulated in the liver and bone marrow resulting in a rather high radiation dose to these organs. The transferrin complex left the vascular compartment with the shortest half-life of any known labelled protein, but was excreted at the rate of only 1 – 3% per day. These findings could not be explained.

^{111}In-chloride (-transferrin) has undergone initial evaluation as a tumour-localizing agent in humans. Hunter and Riccobono [25], using a dose of 3 mCi, performed multiple scans at 24, 48 and 72 h post-administration in over 25 patients with known or highly suspected sites of malignant disease with a wide spectrum of neoplastic disease. There was successful tumour localization at known sites of active disease, and examples were encountered where this uptake was reduced by treatment. In addition, several radiographically unsuspected lesions were suggested by the ^{111}In-chloride scans, some of which were confirmed by follow-up studies. Goodwin and his colleagues [26] have studied ^{111}In injected as the chloride in a dose of 1 mCi in 10 patients with a variety of head and neck tumours. They were successful in localizing the tumour in seven of the patients.

Farrer et al. [27] have reported that in humans, after the administration of 1-2 mCi of ^{111}In-chloride, there is a progressive increase in the ratio of tumour-to-thigh activity with time, a plateau being reached between 48 and 72 h. Untreated malignant head and neck tumours were visualized 24 h post-injection and were best visualized at 48-72 h. In patients who had undergone radiotherapy to lesions, abnormal uptake was observed. A high degree of activity appearing in the bone marrow after 24 h made the visualization of malignant tumour in the thorax and mediastinum difficult.

Merrick and his associates [19] have also studied ^{111}In-chloride as a general tumour-localization agent in humans, and were disappointed with the results. The percentage of positive scans was only 22%, which was inferior to the 60% positive scans obtained with ^{67}Ga-citrate, and the 80% positive scans resulting from the use of ^{111}In-bleomycin.

Zeidler and co-workers [7] investigated the use of ^{111}In as a brain-scanning agent. In this study the radionuclide was bound to the patient's serum before the administration of 5 mCi of activity. The clinical results were similar to those reported from investigations with labelled albumin. Meningiomas formed a strong radioactive focus shortly after administration, as did A-V malformations. Glioblastomas showed a faint and diffused distribution of radioactivity over the tumour region in early scans, but at 24 h an intense radioactive focus was visible. Semi-malignant gliomas showed slow uptake of substance, and were never seen in the early scans; metastases behaved in an unpredictable manner. The substance was also taken up by cerebral-vascular lesions. Twenty-six patients with positive ^{99}Tc^m-pertechnetate scans (12 mCi) were studied. ^{111}In concentrated in 23 of these. The relatively high radiation dose prohibits the administration of large amounts of radioactivity, but the high photon flux from the cascade-decaying radionuclide partially compensates for this. Further development of ^{111}In as a brain-scanning agent may be helpful to the clinician, and it may also find use in head and neck tumours at dose levels of 3-5 mCi per patient.

It should be noted here that there has been successful tumour localization with ^{111}In-labelled antibiotics, particularly ^{111}In-bleomycin. The use of these compounds is described in another paper in these Proceedings.

IV. TUMOUR LOCALIZATION BY PROCESSES OTHER THAN DIRECT TUMOUR UPTAKE

Although not the primary interest of this meeting, it should be noted that the use of compounds labelled with ^{111}In has been helpful in localizing neoplasms which do not take up the radioactive material. Neoplasms appear as an area of decreased activity within the organ. The best examples of this method of tumour localization are the use of labelled indium colloids, usually the phosphate, in liver-spleen scanning, and the use of labelled DTPA in kidney scanning.

^{113}In^m-compounds have also aided in the localization of tumours by means of altered physiology. The most important examples of this are the use of labelled aggregates such as albumin or iron hydroxide in lung scanning, and the use of colloids in lymphatic scanning.

V. SUMMARY

A successful use of $^{113}In^m$-labelled compounds in the localization of tumours by direct uptake has been the use of the $^{113}In^m$-DTPA chelate in brain tumour studies; probably through the mechanism of simple diffusion, a rapid mechanism suitable to the short half-life of the radionuclide. The chloride form of ^{111}In, on direct injection or pre-bound to serum transferrin, seems to be a chemical form showing promise in uptake studies in certain animal tumour models. Its use in general tumour scanning has been limited and thus far results have been inferior to those obtained with ^{67}Ga-citrate. However, from reports of its use as a brain-scanning agent and for tumours in the head and neck region, ^{111}In-chloride in doses of 3 - 5 mCi would seem to warrant further investigation.

REFERENCES

[1] CASTRONOVO, F.P., STERN, H.S., GOODWIN, D.A., Experiences with the ^{113m}Sn-^{113m}In generator, Nucleonics 25 (1967) 64.
[2] COOPER, J.F., WAGNER, H.N., Jr., "Preparation and control of ^{113m}In radiopharmaceuticals", Radiopharmaceuticals from Generator-Produced Radionuclides (Proc. Panel Vienna, 1970), IAEA, Vienna (1971) 83.
[3] TOUYA, J.J., Jr., BONOMI, J.C., FERRANDO, R., TOUYA, E., "Practical problems in the selection of generators and radiopharmaceuticals", Radiopharmaceuticals from Generator-Produced Radionuclides (Proc. Panel Vienna, 1970), IAEA, Vienna (1971) 15.
[4] O'MARA, R.E., SUBRAMANIAN, J.G., McAFEE, J.G., BURGER, C.L., Comparison of ^{113m}In and other short-lived agents for cerebral scanning, J. Nucl. Med. 10 1 (1969) 18.
[5] HINE, G.J., JOHNSON, R.E., Absorbed dose from radionuclides, J. Nucl. Med. 11 7 (1970) 468.
[6] FRENCH, R.J., JOHNSON, P.F., TROTT, N.G., "Dosimetry of indium-113m", Medical Radioisotope Scintigraphy (Proc. Symp. Salzburg, 1968) 1, IAEA, Vienna (1969) 843.
[7] ZEIDLER, U., WEINRICH, W., BRUNNGRABER, C.V., et al., "Indium-111 as a brain-scanning agent", Medical Radioisotope Scintigraphy 1972 (Proc. Symp. Monte Carlo, 1972) 2, IAEA, Vienna (1973) 435.
[8] BURDINE, J.A., Indium-113m radiopharmaceuticals for multipurpose imaging, Radiology 93 (1969) 605.
[9] STERN, H.S., GOODWIN, D.A., SCHEFFEL, V., WAGNER, H.N., Jr., ^{113m}In for blood-pool and brain scanning, Nucleonics 25 3 (1967) 62.
[10] CLEMENTS, J.P., WAGNER, H.N., Jr., STERN, H.S., et al., Indium-113m diethylenetriaminepentaacetic acid (DTPA); a new radiopharmaceutical for brain scanning, Am. J. Roentgenol., Radium Ther. Nucl. Med. 104 (1968) 139.
[11] HAYNIE, T.P., KONIKOWSKI, T., GLENN, H.J., The kinetics of ^{99m}Tc-, ^{113m}In-, and ^{169}Yb-DTPA compounds in brain sarcoma and kidneys of mice, J. Nucl. Med. 13 3 (1972) 205.
[12] ADATEPE, M.H., WELCH, M., EVENS, R.G., Clinical application of the broad spectrum scanning agent — indium 113m, Am. J. Roentgenol., Radium Ther. Nucl. Med. 112 4 (1971) 701.
[13] O'MARA, R.E., MOZLEY, J.M., Current status of brain scanning, Sem. Nucl. Med. 1 1 (1971) 7.
[14] SOLOWAY, A.H., DAVIS, M.A., Survey of radiopharmaceuticals and their current status, J. Pharm. Sc. 63 5 (1974) 647.
[15] MURAYAMA, H., ABE, K., OKAMOTO, S., et al., Diagnostic evaluation of radioisotope in brain tumor, J. Nucl. Med. 11 6 (1970) 348.
[16] CUARON, A., GORDON, F., RODRIGUES, C., Positive imaging of liver tumors by delayed hepatic scintiscanning with acidic ionic indium 113m chloride, Am. J. Roentgenol., Radium Ther. Nucl. Med. 122 2 (1974) 318.
[17] WAGNER, M.S., HUEMER, R.P., SPOLTER, L., BICKERT, C., Radioindium localization in mouse tumors, J. Nucl. Med. 12 6 (1971) 471.
[18] HAYES, R.L., EDWARDS, C.L., "New applications of tumour-localizing radiopharmaceuticals", Medical Radioisotope Scintigraphy 1972 (Proc. Symp. Monte Carlo, 1972) 2, IAEA, Vienna (1973) 531.
[19] MERRICK, M.V., GUNASEKERA, S.W., LAVENDER, J.P., NUNN, A.D., THAKUR, M.L., WILLIAMS, E.D., "The Use of indium-111 for tumour localization", Medical Radioisotope Scintigraphy 1972 (Proc. Symp. Monte Carlo, 1972) 2, IAEA, Vienna (1973) 721.
[20] SERAFINI, A.N., DUNNING, W., CHARYULU, K., WEINSTEIN, M.B., Concentration of ^{111}In-chloride and ^{67}Ga-chloride in the irradiated rat lymphosarcoma, J. Nucl. Med. 12 6 (1971) 464.
[21] HIGASI, T., KANNO, M., KURIHARA, H., MINDELZUN, R.E., A method for the simultaneous measurement of ^{67}Ga, ^{111}In, and ^{75}Se in tumors using semiconductor detector, J. Nucl. Med. 13 8 (1972) 624.

[22] KONIKOWSKI, T., JAHNS, M.F., HAYNIE, T.P., GLENN, H.J., Brain tumor scanning agents in an animal model, J. Nucl. Med. **15** 6 (1974) 508.
[23] KONIKOWSKI, T., HAYNIE, T.P., GLENN, H.J., Kinetics of ^{111}In-bleomycin and ^{111}In-chlorides in mice, J. Nucl. Med. **15** 6 (1974) 508.
[24] KONIKOWSKI, T., GLENN, H.J., HAYNIE, T.P., JAHNS, M.F., An intercomparison of radiopharmaceutical kidney kinetics in the mouse, J. Nucl. Med. **14** 6 (1973) 417.
[25] HUNTER, W.W., RICCOBONO, X.J., Clinical evaluation of ^{111}In for localization of recognized neoplastic disease, J. Nucl. Med. **11** 6 (1970) 329.
[26] GOODWIN, D.A., GOODE, R., BROWN, L., IMBORNONE, C.J., ^{111}In-labeled transferrin for the detection of tumors, Radiology **110** (1971) 175.
[27] FARRER, P.A., SAHA, G.B., SHIBATA, H.N., Evaluation of ^{111}In-transferrin as a tumor scanning agent in humans, J. Nucl. Med. **13** 6 (1972) 429.
[28] LILIEN, D.L., BERGER, H.G., ANDERSON, D.P., BENNETT, L.R., ^{111}In-chloride; a new agent for bone marrow imaging, J. Nucl. Med. **14** 3 (1974) 184.

DISCUSSION

R.L. HAYES: It is interesting to note that you end up with the same tumour-to-non-tumour ratio whether you start with In as a chloride or as a citrate, or whether you start with a strong In chelate. You of course get greatly altered concentrations in the tissues, as most of the activity is actually excreted when you use the chelate form. Obviously, although you get about the same tumour-to-non-tumour ratio, the strong chelate form will be impractical from an economic point of view, because of waste of the radionuclide.

E.H. BELCHER: When we speak about value/cost ratios of Tc and In, it is difficult to give a general or universal statement. The panel meeting on Radiopharmaceuticals from Generator-Produced Radioisotopes in 1971 stated that the cost of using Tc generators averages about twice as much as that of In generators, providing a service based on either one or the other. We know that there are considerable difficulties in using Tc generators in various developing countries, because of the relatively short useful life of the generators, the cost of air transportation etc.

H.J. GLENN: The term "value" is very subjective, depending inter alia on the clinician as well as on other physiological factors.

K. HISADA: You may define the cost performance of a scan image as the "quality" of the image divided by (the radiation exposure to the critical organs times the price of the radiopharmaceutical).

W.H. BEIERWALTES: Dr. Glenn, I wonder if you could remind us of the reasons why we had inferior results when we used methylethylketone extracts from our reactor-produced Tc, whereas we have no trouble after shifting to generator-produced Tc.

H.J. GLENN: There was a time when the pertechnetate available by the MEK extraction procedures probably had impurities in the form of aldols and other compounds like oxidizing agents. There has also been a time when there was a quality control problem with the polyphosphates. Similar problems have been encountered less frequently with pyrophosphate, which is a definite compound, and with EHDP which is also a definite compound, although the labelling of EHDP may be a little more irregular.

K. HISADA: As some of our studies have only been published in Japan, I should like to tell about some of our results.

The tumour affinity of Yb-citrate, Ga-citrate, In-citrate and ^{57}Co-bleomycin has been studied in Yoshida sarcoma-bearing rats and also tested in rats with inflammation induced by croton oil injections. All the compounds had a relatively strong affinity for the inflamed tissue. The concentration ratios of the compounds between tumour and inflammatory tissue were 2.1 to 2.2 at 24 hours after intravenous injection. These figures were nearly the same for all three or four compounds. In addition, there was no great difference in the uptake in the tumour tissue between the compounds, but great differences were found in the biological properties of the compounds, as observed in the retention value in the blood and the uptake rate in the bones. Yb-citrate was rapidly cleared from the blood and was taken up by the bones. On the other hand, ^{111}In-citrate was slowly cleared from the blood and only slightly taken up by the bones. ^{67}Ga-citrate gave intermediate results.

In some other short-time experiments we studied the uptake rate of the different compounds. ^{169}Yb already reached 70% of the final level within 10 min after injection. At this time ^{67}Ga had reached 50% of the final value, and ^{111}In had already reached the final level. On the basis of these facts we look upon the chemical bond of these elements not as a chelation, but as an ionic bond.

Finally, we compared Co-bleomycin, Tc-bleomycin, and In-bleomycin as tumour-scanning agents. The studies were carried out in rats bearing Yoshida sarcoma and in rats with inflamed tissues (as already mentioned). Co-bleomycin had excellent features as a tumour-scanning agent, but Tc-bleomycin and In-bleomycin gave poor results. When we used Co-bleomycin and Tc-bleomycin in patients, we found a high positive tumour detection with Co-bleomycin, but nearly all cases were negative when we used Tc-bleomycin.

T. MUNKNER: Your studies have shown that your major uptake takes place in the first minutes after the injection. I wonder if anyone here could tell whether there is a difference in the per cent uptake in the tumour tissues between a rapid injection (a bolus) and a slower intravenous injection.

D. COMAR: In studies a couple of years ago at Orsay, intravenous injections and intraperitoneal injections were compared, Co-bleomycin being injected into tumour-bearing mice. The intraperitoneal route of administration gave the higher uptake in the tumour, and it was argued that it is better when the product arrives slowly to the tumour. This obviates a situation where you might saturate the tumour tissue with the injected compound. The studies were carried out with carrier.

H.J. GLENN: I don't know exactly what the effect is of the type of injection. We have made very careful studies on the relationship between renal clearance and tumour uptake for something like 15 different compounds. It is clear that renal clearance has a striking influence on tumour uptake.

W.H. BEIERWALTES: There is no question that there are certain elements that will concentrate to a higher amount in tissues if the concentration in the blood is higher. There are, however, also instances where the concentration constantly increases long after the blood level has fallen practically to zero. We have tried to inject ^{75}Se-methionine intra-arterially as a bolus at the level of the parathyroid glands, and it resulted in a rather striking increase of ^{75}Se methionine in the parathyroids. Somehow, more complicated results were found when we studied the uptake of the ^{125}I- or ^{131}I-chloroquine analogue. This compound is excreted to a large extent by the liver, through the bile and out into the bowel. Within a matter of two or three days most of the radioactivity has gone from the bile and even from the bowel. Even if the blood activity at this time has fallen strikingly, the concentration of radioactivity in the choroid of the eye rises progressively in the human for about 14 days (it peaks between 14 and 21 days). One possible explanation of this phenomenon may be that the compound concentrating in the choroid of the eye may not be the original chloroquine analogue, but a metabolite, and as the compound is metabolized in the liver, more of the compound might be available to the choroid of the eye. It is also worth mentioning that you have practically no ^{131}I-19-iodocholesterol uptake in the adrenal cortex during a dexamethazone suppression test, but if you suddenly stop the suppression and make images of the adrenals in the following days, you will find a striking uptake of radioactivity in the adrenocortex within three days even if you haven't given a second dose to the patient. From these examples I think it evident that you should not only study the disappearance from the blood and the tissue concentrations of radionuclide-labelled compounds in the first four or eight hours or so, but you should investigate the changing picture of increasing uptakes over a period of maybe two or three weeks.

E.H. BELCHER: It is reasonable to stress that the best radionuclide for one situation may not necessarily be the best for another or all situations. The high gamma-ray energy of ^{113}In, for example, is often cited as a disadvantage as compared with ^{99m}Tc, but for the detection of deeply situated tumours, for example in brain, it may not necessarily be a bad thing, if we think in terms of photon attenuation in tissues.

H.J. GLENN: I have had the opportunity to observe the use of indium colloid for liver scintigraphy followed by the use of Tc-sulphur colloid. The largest difference that I noted was in the lateral view of the liver, where the In images appeared to be better than the Tc images because of the difference in radiation absorption. The spleen may be seen "shining" through in the lateral scan also because of the more energetic radiation of ^{113}In^m.

R.L. HAYES: The matter of attenuation is a rather imporant one but when it comes to comparison of various radionuclides, instrumentation must necessarily enter into the picture. For example, if you are going to compare ^{99}Tc^m-bleomycin with ^{67}Ga in the same patient, or in a series of similar patients, the use of a camera necessarily biases the comparison of bleomycin with ^{67}Ga towards the ^{99}Tc^m-labelled compound. If you use a rectilinear scanner, you could very well bias things in favour of ^{67}Ga. Maybe it is more important to utilize the best instrumentation for each radionuclide, when it is available.

H.J. GLENN: In the studies I just referred to a gamma camera with a straight bore high-energy collimator was used for the In and a straight bore low-energy collimator for the Tc.

W.H. BEIERWALTES: It is indeed very important to state that there is not just one energy that should be used in nuclear medicine, namely 140 keV ± 20 keV. There are a number of examples which show that Tc is not good enough for all kinds of studies. It is enough to mention that substernal goitres and very large goitres are better studied with ^{131}I than with ^{99}Tc^m. The long-term answer may be that we want some high-energy radionuclides for certain specific purposes. It may easily become a trap when we only consider the 140-keV gamma as very fit for the Anger camera and then conclude that this is the future tool of nuclear medicine.

LABELLED ANTIBIOTICS AS TUMOUR-LOCALIZING AGENTS

D.M. TAYLOR, V.R. McCREADY
Departments of Radiopharmacology
and Nuclear Medicine,
Institute of Cancer Research
and Royal Marsden Hospital,
Sutton, Surrey,
United Kingdom

Abstract

LABELLED ANTIBIOTICS AS TUMOUR-LOCALIZING AGENTS.
The published results of clinical and experimental studies of labelled bleomycins and tetracyclines are reviewed. None of the labelled antibiotics yet studied show anything approaching absolute tumour specificity. Clinical trials suggest that ^{57}Co-bleomycin is superior to either ^{111}In- or ^{99}Tc^m-bleomycin and that it may possess some advantages over ^{67}Ga-citrate in respect of lower uptake in the abdomen and, possibly, lower uptakes in benign and inflammatory lesions. Radioiodine-labelled or ^{99}Tc^m-labelled tetracyclines appear to be of little value in tumour localization.

INTRODUCTION

Interest in the possible use of labelled antibiotics for tumour localization stems from observations that tetracyclines [1–3] and, later, bleomycins [4,5] showed some degree of concentration in tumours. Since 1960 several attempts have been made to use radioiodine-labelled tetracyline or, more recently, ^{99}Tc^m-labelled tetracycline for tumour visualization without any real degree of success [6–8]. However, since the original reports by Nouel and his colleagues [9] that bleomycin labelled with ^{57}Co could be used for tumour localization and for the assessment of the spread of malignant disease, there has been widespread interest in the potential value of bleomycins labelled with a number of different radionuclides.

This paper presents a review of clinical and experimental studies of the value of radionuclide-labelled bleomycins and tetracyclines in tumour localization and staging.

LABELLED BLEOMYCINS

Chemistry of bleomycin

Bleomycin, which was first isolated by Umezawa and co-workers [10], is produced by the fungus *Streptomyces verticullus*: it is not a single compound but consists of a mixture of polypeptides with a common base structure but differing terminal amine moieties. It is cytotoxic and antimicrobial and is used clinically in the treatment of some tumours, particularly those of epidermal origin.

Bleomycin was originally isolated as a copper complex [10] and studies by Renault et al. [11] have shown that a number of di- and tri-valent metal ions form complexes with bleomycin. Most of these complexes are of low stability and Renault concluded that only copper and cobalt complexes of bleomycin would be sufficiently stable for biological use [11].

TABLE I. COMPARISON OF TUMOUR-TO-TISSUE RATIOS 24 HOURS AFTER ADMINISTRATION OF LABELLED BLEOMYCINS TO TUMOUR-BEARING ANIMALS

Nuclide	Tumour	Animal	Tumour-to-blood	Tumour-to-liver	Tumour-to-muscle	Ref.
^{57}Co	Ehrlich carcinoma (solid)	Mouse	72.0	2.9	24.0	
^{111}In			3.4	0.4	3.9	[12]
^{67}Ga			1.8	0.2	8.4	
^{57}Co	Renal carcinoma	Rat	30.0	2.7	52.5	
^{62}Zn			2.8	0.6	5.3	[13]
^{111}In			1.3	0.6	6.7	
^{111}In	Rib 5-sarcoma	Rat	11.5	–	–	
^{111}In	Walker	Rat	5.1	–	–	[17]
^{67}Ga	Walker	Rat	3.7	–	–	
^{111}In	Lewis lung	Mouse	4.4	1.7	3.1	
^{111}In	B-16 melanoma	Mouse	5.8	1.3	7.8	[18]
^{111}In	Ridgeway osteosarcoma	Mouse	2.6	1.3	2.9	
^{57}Co	?	Mouse	18.0	1.0	–	[23]
^{57}Co	Mammary adenocarcinoma	Rat	16.8	–	2.7	[16]
^{67}Cu			1.3	–	1.8	
^{99}Tc^m	KHJJ carcinoma	Mouse	2.8^a	1.6^a	15.5^a	[19]
^{99}Tc^m	Fibrosarcoma	Mouse	10^a	1^a	–	[44]
^{140}La	Sarcoma-180	Mouse	4.8	0.03	10.0	
^{153}Sm			16	0.2	10.5	[21]
^{169}Yb			219	3.0	10.8	

^{a} 6 hours after injection.

Experimental studies with labelled bleomycins

Studies of tissue distribution and tumour uptake in experimental animals have been reported using bleomycin labelled with the following nuclides: ^{57}Co [12, 13]; ^{67}Cu [14–16]; ^{62}Zn [13]; ^{111}In [12, 13, 17, 18]; ^{67}Ga [12]; ^{99}Tc^m [17, 19]; ^{51}Cr [20]; ^{239}Np, ^{237}U, ^{140}La, ^{153}Sm and ^{169}Yb [21]:

Comparative studies with ^{57}Co-, ^{111}In- and ^{67}Ga-bleomycin [12] or with ^{57}Co-, ^{62}Zn- and ^{111}In-bleomycin [13] in tumour-bearing rats have shown that, although in each study the initial tumour uptake was similar for the three complexes studied, the more rapid plasma and tissue clearance of ^{57}Co resulted in much greater tumour-to-blood and tumour-to-liver ratios for this nuclide, at 24 hours post-injection, than for ^{111}In, ^{67}Ga or ^{62}Zn. In other studies in rats bearing implanted tumours, Thakur et al. [17] found that the tumour-to-blood ratio was higher for ^{111}In-bleomycin than for either ^{67}Ga- or ^{99}Tc^m-bleomycin. The uptake of ^{99}Tc^m-bleomycin in mice bearing an implanted carcinoma has been studied by Lin et al. [19] who found that the tumour-to-blood ratio did not exceed 3 at 6 hours after injection.

From theoretical considerations and from the work of Renault et al. [11], radioactive copper-bleomycin complexes might be expected to show a similar behaviour to that of ^{57}Co-bleomycin. Some experimental studies using ^{67}Cu- or ^{64}Cu-labelled bleomycin have been reported: Coates et al. [14] compared the clearance of ^{57}Co-, ^{111}In- and ^{67}Cu-bleomycin from the plasma of dogs and found that the clearance patterns were similar for ^{57}Co and ^{67}Cu; no increase in liver radioactivity was observed suggesting that the ^{67}Cu-bleomycin complex did not dissociate. In studies in tumour-bearing animals, Hall and O'Mara [15] found that the tumour-to-muscle ratio of ^{67}Cu-bleomycin in a hepatoma was approximately double that of ^{111}In-bleomycin; however, the observations of Grove et al. [12] and Taylor and Cottrall [13] suggest that the corresponding ratios for ^{57}Co-bleomycin would be more than eight times greater than those for ^{111}In-bleomycin. Studies with ^{64}Cu-bleomycin in a mouse mammary tumour by Eckelman et al. [16] showed a tumour-to-blood ratio of 1.30 for ^{64}Cu compared with 16.8 for ^{57}Co-bleomycin; tumour-to-muscle ratios were 1.83 for ^{64}Cu and 2.68 for ^{57}Co.

The tumour-to-blood, tumour-to-muscle and tumour-to-liver ratios observed with the different bleomycin complexes studied are summarized in Table I. These data indicate that the ^{57}Co- and ^{111}In-complexes show the highest ratios; ^{67}Cu may prove to be an improvement on ^{111}In though ^{67}Cu is difficult and expensive to produce. Ytterbium-169-bleomycin shows a very high tumour-to-blood ratio and further studies with this complex may prove interesting.

The long physical half-life of ^{57}Co makes this nuclide undesirable for routine use. In man, 70–90% of administered ^{57}Co-bleomycin is excreted in the urine in 24 hours and, consequently, the patient radiation dose is relatively low. However, the large amount of the nuclide excreted in the urine presents problems in respect of disposal, contamination risks and staff exposure. There is strong evidence from both clinical and animal studies that the ^{111}In-bleomycin complex is not stable in vivo and that the indium is displaced from the bleomycin and becomes bound by transferrin and other plasma and tissue proteins [13, 17]. Studies of the subcellular distribution of ^{111}In and ^{57}Co in a rat tumour after simultaneous administration of the two nuclides as their chlorides or as the bleomycin complexes [13] showed that ^{111}In was located predominantly in the lysosomal fraction in each case, whereas the ^{57}Co was associated mainly with the nuclear fraction after administration as the bleomycin complex, and in the soluble fraction after administration of the chloride. Similar observations have been made in a mouse tumour after administration of ^{57}Co-bleomycin, or Co-BLM-^{14}C, and ^{57}CoCl$_2$ [22]. These results suggest that the ^{57}Co-bleomcyin complex does not dissociate in vivo, a view which is supported by the observations that in man ^{57}Co-bleomcyin is excreted unchanged in the urine [23].

The mechanism of uptake of labelled bleomycins by tumours is not yet understood, neither is the relationship of the uptake to cell proliferation and metabolic activity. In a small series of cases Cheguillaume and his colleagues [24] observed an inverse correlation between ^{57}Co-bleomycin uptake and tumour doubling time in human pulmonary tumours.

Clinical experience with labelled bleomycin

Clinical studies using ^{57}Co-, ^{111}In- or ^{99}Tc^m-bleomycin have been reported from a number of centres.

Cobalt-57-bleomycin

This complex has been studied most extensively by Nouel and his associates [9, 23, 25] who have recently reviewed their experience with 1000 patients. Studies in smaller groups of patients have been reported by Nakano et al. [26] and by other groups [12, 27–29]. Some of

TABLE II. COMPARISON OF RESULTS OF COBALT-57-BLEOMYCIN SCANS (24–72 h)

Tumour site	Number of positive scans			
	Nouel et al. [25]		Nakano et al. [26]	
Lung	267/273	(98%)	21/24	(87%)
Breast	22/45	(49%)	5/9	(55%)
Pancreas	5/7	(71%)	3/4	(75%)
Oesophagus	16/16	(100%)	0/1	(0%)
Stomach	3/4	(75%)	0/1	(0%)
Liver — hepatocarcinoma	–		6/12	(50%)
cholangiosarcoma	–		2/2	(100%)
metastatic Ca	–		0/1	(0%)
Brain — glioblastoma + astrocytoma-GR.4	44/45	(98%)	–	
meningioma	3/12	(25%)	–	
oligodendroglioma	5/5	(100%)	–	
vascular accidents	0/15	(0%)	–	
benign lesions	0/22	(0%)	–	
Benign conditions				
pulmonary tuberculosis	13/29	(45%)	1/2	(50%)
others	8/84	(9%)	0/3	(0%)
Total	346/557	(62%)	38/59	(64%)

the results of Nouel et al. [25] and Nakano et al. [26] are summarized in Table II which compares the number of positive scans observed in different tumour sites; when the numbers of cases are large enough to allow meaningful comparison there is good agreement between the two groups. In their studies, Nouel et al. [25] administered 1 to 1.5 mCi ^{57}Co-bleomycin (15 mg bleomycin) and scanned at 24 and 72 hours later. Nakano et al. [26] and Suzuki et al. [29] used a dose of 500 μCi ^{57}Co-bleomycin (4 to 5 mg bleomycin) and scanned at 24 hours.

High proportions of positive scans were observed in tumours of lung, brain, stomach, oesophagus and pancreas. Nouel et al. [25] studied 46 cases of Hodgkin's disease and found that superficial lymphadenopathies were always visualized but that lumbo-aortal or splenic involvement was not demonstrated. Skeletal or cerebral metastases were discovered in two cases. Detection of metastatic lesions has been demonstrated in a high proportion of patients with primary tumours at different sites.

An important part of the evaluation of new radiopharmaceuticals is their comparison with established agents. Watanabe et al. [28] have compared ^{57}Co-bleomycin and ^{67}Ga-citrate in 46 patients, including 19 lung tumours, 10 malignant lymphomas, 2 oesophageal and 5 intraperitoneal tumours, and concluded that ^{57}Co-bleomycin was not superior to ^{67}Ga-citrate. However, Grove et al. [12] conclude from studies in 15 patients that ^{57}Co-bleomycin may show a higher tumour detection efficiency than ^{67}Ga, especially for adenocarcinoma. Nouel et al. [25] have reported that ^{57}Co-bleomycin shows higher tumour-to-normal tissue ratios than ^{67}Ga in lung tumour and in cerebral metastases. Like ^{67}Ga, ^{57}Co-bleomycin localizes in some non-malignant lesions, such as abscesses, and in acute inflammatory lesions [25, 26].

On the basis of their extensive experience Nouel and his associates [25] conclude that ^{57}Co-bleomycin is advantageous in the investigation of primary tumours, particularly in lung and brain,

TABLE III. COMPARISON OF RESULTS OF INDIUM-111-BLEOMYCIN SCANS (24 h)

Tumour site	Number of positive scans			
	Lilien et al. [32]		Yeh et al. [35]	
Lung	38/40	(95%)	9/16	(56%)
Breast	15/21	(71%)	5/8	(62%)
Colon	8/10	(80%)	—	
Lymphoma	59/75	(79%)	—	
Head and neck	7/10	(70%)	4/4	(100%)
Melanoma	14/14	(100%)	15/17	(88%)
Ovary	13/15	(88%)	—	
Miscellaneous sarcoma	9/10	(90%)	6/8	(75%)
Total	184/195	(94%)	39/53	(74%)

but that it is especially valuable in assessing metastatic spread when the site of the primary tumour is unknown; in the brain it is also useful in distinguishing gliomas from metastatic lesions. The other reported experience does not contradict these conclusions.

Indium-111-bleomycin

The use of ^{111}In as a label for bleomycin was first studied by Merrick et al. [30, 31] who found that ^{111}In-bleomycin consistently localized in both primary and secondary tumours, especially in carcinoma of the bronchus. Uptake in pelvic lesions was also observed. Lilien et al. [32, 33] have carried out 232 studies in 108 patients and found that in 198 scans (84%) there was good correlation with known sites of disease. False positives were observed in 22 studies (10%) and false negatives in 15 (10%); some false positives occurred in inflammatory lesions and diffuse pulmonary uptake was seen in patients treated with alkylating agents. Results for particular tumour sites are summarized in Table III. Goodwin et al. [34] have reported positive scans in 20 out of 29 patients with proven malignancy (69%); 4 of the 9 false negatives had lesions less than 3 cm in diameter and 2 had received prior radiotherapy. Studies in 96 patients have been reported briefly by Yeh et al. [35] who found positive uptake of ^{111}In-bleomycin in 71 cases (74%); some of the findings for individual sites are summarized in Table III. Detectable localization of ^{111}In-bleomycin in 80 out of 101 patients (80%) has also been reported by Verma et al. [36].

Comparative studies with ^{111}In-bleomycin and ^{67}Ga-citrate have been reported by several groups of workers. In 9 patients investigated with ^{111}In-bleomycin, ^{57}Co-bleomycin and ^{67}Ga-citrate, Grove et al. [12] observed positive scans with ^{111}In in 2 out of 9 patients (22%) compared with 7 out of 9 (78%) for ^{57}Co-bleomycin and 5 out of 9 (55%) for ^{67}Ga-citrate. Oyamada et al. [37] also report that scan quality and tumour detection efficiency were better with ^{67}Ga-citrate than with ^{111}In-bleomycin. Similar findings have been reported by Paterson et al. [38] in 20 patients with Hodgkin's disease who were investigated with both ^{67}Ga-citrate and ^{111}In-bleomycin.

The low uptake of ^{111}In-bleomycin in abdominal organs facilitates the demonstrations of pelvic lesions. However, the generally poorer quality of the scans obtained suggests that for most other tumour sites this complex offers no advantages of ^{67}Ga-citrate or ^{57}Co-bleomycin as a tumour-localizing agent.

TABLE IV. COMPARISON OF TECHNETIUM-99m-BLEOMYCIN (0.25–1 h) AND GALLIUM-67-CITRATE (72 h) SCANS [39]

Tumour site	Number of positive scans			
	$^{99}Tc^m$-bleomycin		^{67}Ga-citrate	
Orbit	12/12	(100%)	1/4	(25%)
Head and neck	40/48	(83%)	29/31	(94%)
Thyroid	41/43	(95%)	5/24	(21%)
Lung	40/52	(77%)	34/43	(79%)
Oesophagus	6/11	(55%)	6/9	(67%)
Liver or pancreas	10/17	(59%)	9/16	(56%)
G.I. tract	5/7	(71%)	2/7	(29%)
Pelvis	0/2	(0%)	0/2	(0%)
Lymph node	9/13	(69%)	13/13	(100%)
Total	193/239	(81%)	114/176	(65%)

Technetium-99m-bleomycin

The combination of short physical half-life, good scanning characteristics and ready availability of $^{99}Tc^m$ would make a $^{99}Tc^m$-bleomycin complex an attractive radiopharmaceutical for tumour imaging, and clinical and experimental studies with this complex have been made by several groups. Mori et al. [39] have reported studies in 390 patients with a $^{99}Tc^m$-bleomycin complex prepared by a stannous chloride — ascorbic acid technique in which 3 to 5 mCi were administered and the patient was scanned 15 to 60 min later. Some of the results of this study are summarized in Table IV together with the results of ^{67}Ga-citrate scans on the same patients. The overall efficiency of tumour localization with $^{99}Tc^m$-bleomycin was 81% which is similar to that observed with ^{111}In- and ^{57}Co-bleomycins. Adenocarcinomas were well visualized but malignant lymphomas were poorly demonstrated and high blood and abdominal radioactivity were present at the time of scanning. The incidence of positive uptake in inflammatory and benign lesions was 17% with the $^{99}Tc^m$ complex compared with 45% for ^{67}Ga-citrate. Lin et al. [19], using a similarly prepared $^{99}Tc^m$-bleomycin complex, observed positive localization in 9 out of 18 patients with biopsy-proven malignant lesions, and in general the lesions visualized were either large or superficially located. High background was observed in the region of the bladder, kidneys, liver, stomach and heart.

Technetium-99m-bleomycin complex prepared by an electrolytic method has been studied by Oyamada et al.[37] who found that visualization efficiency was too low for clinical use. This latter observation raises doubts concerning the chemical identity of $^{99}Tc^m$-bleomycin complexes prepared by different methods and further work is needed to clarify this matter before extensive further clinical investigation of this complex is undertaken.

LABELLED TETRACYCLINES

In contrast to the studies with labelled bleomycins, little attention has been concentrated on radionuclide-labelled tetracyclines as potential tumour-localizing agents.

The retention of tetracyclines in tumours was first described by Rall and his associates [1–3] in 1957, and some investigations of the potential value of ^{131}I-labelled tetracycline for tumour localization were reported in 1960 [6, 7]. At that time tumour uptake was relatively low and doubts were expressed concerning the stability of iodotetracyclines in vivo [7]. More recent chemical studies of tetracycline uptake in tumours suggest that the greatest uptake of the drug is observed in areas of necrosis, especially early necrosis [41].

Interest in labelled tetracyclines as potential tumour-scanning agents has been re-awakened by reports of studies with $^{99}Tc^m$-tetracyclines and of further studies with ^{131}I-tetracyclines.

Chauncey et al. [42] have developed high efficiency procedures for labelling tetracycline and chlorotetracycline with ^{131}I and shown that in rat hepatoma, tumour-to-normal tissue ratios of 20:1 may be obtained with ^{131}I-tetracycline; further animal work is needed to confirm these findings in other types of tumour before clinical investigation is undertaken.

In other recent reports, Holman et al. [8, 43] have studied the uptake of $^{99}Tc^m$-tetracycline in tumours in mice, rats and rabbits and in 18 patients with proven malignant disease. In the mouse and rat tumours at 24 hours after injection the tumour-to-blood and tumour-to-muscle ratios for $^{99}Tc^m$-tetracycline were greater than those for ^{67}Ga-citrate, although the concentration of ^{67}Ga in the tumour was greater than that of $^{99}Tc^m$. In three rabbits the tumours (V2 carcinoma) could be clearly visualized by scanning at 24 hours after injection.

In the clinical studies, tumours were clearly demonstrated in 14 out of 18 cases (78%) and the scan accurately delineated the number and extent of the lesions, particularly in untreated patients. Tumour detection was particularly good in the chest where 85% of lesions were detected. Detection efficiency was poor in patients who had received radio- or chemotherapy. A number of disadvantages of $^{99}Tc^m$-tetracycline were observed including high gastro-intestinal and genito-urinary activity in man. Like ^{67}Ga-citrate, $^{99}Tc^m$-tetracycline is unable to distinguish tumour from infarct, infection or necrosis.

On the basis of the investigations reported up to the present, labelled tetracyclines appear to have little general value as tumour-localizing agents.

CONCLUSIONS

The studies reviewed here suggest that, although they are not tumour-specific, radionuclide-labelled bleomycins may have a useful role to play in tumour localization and staging, and in the differential diagnosis of certain types of lesion. Bleomycin complexes appear to offer some advantages over ^{67}Ga-citrate in respect of lower uptake in the abdominal cavity and, possibly, in inflammatory and other benign conditions.

Of the three labelled bleomycins subjected to clinical trial, ^{57}Co-bleomycin, which is stable in vivo, appears to yield the most consistent results, although the long half-life of ^{57}Co is a practical disadvantage. Radiation dose estimates for ^{57}Co-bleomycin and ^{111}In-bleomycin are 0.03 and 0.15 rad/mCi for the whole body and 0.45 and 1.4 rad/mCi respectively for the bladder wall, the critical organ. The whole body dose from ^{67}Ga is estimated to be 0.26 rad/mCi and that to the kidney 1.2 rad/mCi [12, 40].

Indium-111-bleomycin is commercially available in the United States, but ^{57}Co- and $^{99}Tc^m$-bleomycins must be prepared by the user. At the present time the costs of investigations with ^{57}Co- or ^{111}In-bleomycins are comparable with those of ^{67}Ga-citrate.

There is scope for much further work in the evaluation of $^{99}Tc^m$-bleomycin and in the search for other radionuclide labels for bleomycin.

In contrast to the labelled bleomycins, labelled tetracyclines appear to have little value as tumour-localizing agents. Attempts to prepare potential tumour-localizing radiopharmaceuticals from other cytotoxic antibiotics, such as adriamycin, might yield some interesting results. However, on the evidence available at the present time it seems unlikely that such radiopharmaceuticals would exhibit very markedly greater tumour-specificity than that of currently available agents.

REFERENCES

[1] RALL, D.P., LOO, T.L., LANE, M., KELLY, M.G., J. Natl. Cancer Inst. **19** (1957) 79.
[2] MILCH, R.A., RALL, D.P., TOBIE, J.E., J. Natl. Cancer Inst. **19** (1957) 87.
[3] TITUS, E.D., LOO, T.L., RALL, D.P., Antibiot. Ann. (1957–58) 949.
[4] UMEZAWA, H., TAKEUCHI, T., HORI, S., SAWA, T., ISHIZUKA, M., ICHIKAWA, T., KOMAI, T., J. Antibiot. **25** (1972) 409.
[5] HAYAKAWA, T., USHIO, T., MOGAMI, H., HORIBATA, K., Eur. J. Cancer **10** (1974) 137.
[6] DUNN, A.L., ESKELSON, C.D., McLEAY, J.F., OGBORN, R.E., WALSKE, B.R., Proc. Soc. Exp. Biol. Med. **104** (1960) 12.
[7] HLAVKA, J.J., BUYSKE, D.A., Nature (London) **186** (1960) 1064.
[8] HOLMAN, B.L., KAPLAN, W.D., DEWANJEE, M.K., FLIEGEL, C.P., DAVIS, M.A., SKARIN, A.T., ROSENTHAL, D.S., CHAFFEY, J., Radiobiology **112** (1974) 147.
[9] NOUEL, J.-P., RENAULT, H., ROBERT, J., JEANNE, C., WICENT, L., La Nouvelle Presse Medicale **1** (1972) 25.
[10] UMEZAWA, H., Biomedicine **18** (1973) 459.
[11] RENAULT, H., HENRY, R., RAPIN, J., HEGELSIPPE, M., in Radiopharmaceuticals and Labelled Compounds (Proc. Symp. Copenhagen, 1973) **2**, IAEA, Vienna (1973) 195.
[12] GROVE, R.B., ECKELMAN, W.C., REBA, R.C., J. Nucl. Med. **14** (1973) 917.
[13] TAYLOR, D.M., COTTRALL, M.F., Radiopharmaceuticals (SUBRAMANIAN, G., RHODES, B.A., COOPER, J.F., SODD, V.J., Eds), (Proc. Int. Symp. on Radiopharmaceuticals, Atlanta, Ga., 1974), Society of Nuclear Medicine, New York (1975) 458.
[14] COATES, G., ASPIN, N., WONG, P.Y., WOOD, D.E., J. Nucl. Med. **15** (1974) 484.
[15] HALL, J.N., O'MARA, R.E., J. Nucl. Med. **15** (1974) 498.
[16] ECKELMAN, W.C., REBA, R.C., KUBOTA, H., STEVENSON, J., J. Nucl. Med. **15** (1974) 489.
[17] THAKUR, M.L., MERRICK, M.V., GUNASEKERA, S.W., in Radiopharmaceuticals and Labelled Compounds (Proc. Symp. Copenhagen, 1973) **2**, IAEA, Vienna (1973) 183.
[18] ROBBINS, P.J., SILBERSTEIN, E.B., FORTMAN, D.L., J. Nucl. Med. **15** (1974) 273.
[19] LIN, M.S., GOODWIN, D.A., KRUSE, S.L., J. Nucl. Med. **15** (1974) 338.
[20] ORII, H., Jap. J. Clin. Radiol. **18** (1973) 211.
[21] RAYADU, G.V.S., FORDHAM, E.W., RAMCHANDRAN, P.C., FRIEDMAN, A.M., SULLIVAN, J., J. Nucl. Med. **15** (1974) 526.
[22] KONO, A., KOJIMA, M., MAEDA, T., Jap. J. Clin. Radiol. **18** (1973) 195.
[23] MAMO, L., NOUEL, J.-P., CHAI, N., HOUDART, R., J. Neurosurg. **39** (1973) 735.
[24] CHEGUILLAUME, J., MINIER, J., TUCHAIS, C., TUCHAIS, E., LENK, S., OURY, M., Private communication (1974).
[25] NOUEL, J.-P., ROBERT, J., BERTRAND, A., WITZ, H., DELORME, G., RENAULT, H., MAMO, L., Proc. First World Congr. Nucl. Med., Tokyo (1974) 129.
[26] NAKANO, S., HASEGAWA, Y., ISHIGAMI, S., Proc. First World Congr. Nucl. Med., Tokyo (1974) 700.
[27] TAKASU, A., NAGOSHI, Y., USNI, I., YAMAGUCHI, O., Proc. First World Congr. Nucl. Med., Tokyo (1974) 706.
[28] WATANABE, K., KAWAHIRA, K., KAMOI, I., MORITA, K., MATSUURA, K., Proc. First World Congr. Nucl. Med., Tokyo (1974) 937.
[29] SUZUKI, Y., HISADA, K., HIRAKI, T., ANDO, A., Radiology **113** (1974) 139.
[30] MERRICK, M.V., GUNASEKERA, S.W., LAVENDER, J.P., NUNN, A.D., THAKUR, M.L., WILLIAMS, E.D., in Medical Radioisotope Scintigraphy 1972 (Proc. Symp. Monte Carlo, 1972) **2**, IAEA, Vienna (1973) 721.
[31] THAKUR, M.L., MERRICK, M.V., GUNASEKERA, S.W., LAVENDER, J.P., in Bleomycin in the Treatment of Malignant Disease, Lundbeck, London (1974) p.19.
[32] LILIEN, D.L., JONES, S.E., O'MARA, R.E., SALMON, S.E., DURIE, B.G., J. Nucl. Med. **15** (1974) 512.
[33] JONES, S.E., LILIEN, D.L., O'MARA, R.E., DURIE, B.G.M., SALMON, S.E., Private communication (1974).
[34] GOODWIN, D.A., SUNDBERG, M.W., DIAMANTE, C.I., MEARES, C.F., Radiopharmaceuticals (SUBRAMANIAN, G., RHODES, B.A., COOPER, J.F., SODD, V.J., Eds), (Proc. Int. Symp. on Radiopharmaceuticals, Atlanta, Ga., 1974), Society of Nuclear Medicine, New York (1975) 80.
[35] YEH, S.D.J., GLANDO, R., YOUNG, C.W., BENUA, R.S., Proc. First World Congr. Nucl. Med., Tokyo (1974) 928.
[36] VERMA, R.C., BENNETT, L.R., TOUYA, J.J., MORTON, D.C., WITT, E., J. Nucl. Med. **14** (1973) 641.
[37] OYAMADA, H., ISHIBASHI, H., ORII, H., Proc. First World Congr. Nucl. Med., Tokyo (1974) 707.
[38] PATERSON, A.H.G., TAYLOR, D.M., McCREADY, V.R., Br. J. Radiol. **48** (1975) 832.
[39] MORI, T., ODORI, T., SAKAMOTO, T., HAMAMOTO, K., ONOYAMA, Y., TORIZUKA, K., Proc. First World Congr. Nucl. Med., Tokyo (1974) 703.

[40] SAUNDERS, M.G., TAYLOR, D.M., TROTT, N.G., Br. J. Radiol. **46** (1973) 456.
[41] WINKELMAN, J., GORTEIN, F., Experientia **27** (1971) 309.
[42] CHAUNCEY, D.M., HALPERN, S.E., ALAZRAKI, N.P., J. Nucl. Med. **15** (1974) 483.
[43] DEWANJEE, M.K., FLIEGEL, C.P., HOLMAN, B.L., DAVIS, M.A., J. Nucl. Med. **14** (1973) 624.
[44] MORI, T., HAKAM OTO, K., TORIZUKA, K., Jap. J. Clin. Radiol. **18** (1973) 201.

DISCUSSION

H. LANGHAMMER: Labelled bleomycin, particularly ^{57}Co-bleomycin, has two advantages over ^{67}Ga: (1) Normally radioactivity can only be seen in the kidneys and the bladder, and that means that the detectability of neoplasmas in the abdominal area is better with this compound; (2) Bleomycin has a significantly lower uptake in benign lesions, such as inflammations and infections, and it does show an increased accumulation in adenocarcinomas.

K. HISADA: ^{57}Co-bleomycin is now commercially available in Japan. Bleomycin itself has no special affinity for tumours, but labelled with ^{57}Co or with rhodium it does have tumour affinity. It is interesting to note that cobalt bleomycin hasn't shown any anti-tumour action. It seems as though tumour affinity of labelled bleomycin has nothing to do with anti-tumour action.

D. COMAR: There may be differences between different lots of ^{57}Co-bleomycin. It is important to select the bleomycin preparations, and for the moment Dr. Nouel has selected some lots which seem to be better than others.

V.R. McCREADY: Does Dr. Nouel screen the bleomycin before he starts to use it for clinical studies?

D. COMAR: I guess yes, but I don't know anything about the procedures.

V.R. McCREADY: Dr. Comar, have you any idea why Co-labelled bleomycin concentrates in a tumour? Is the compound split for instance into bleomycin and cobalt?

D. COMAR: The dissociation constant for Co-bleomycin is very small. All the cobalt excreted in the urine is in the form of ^{57}Co-bleomycin, and this was a big problem to Dr. Nouel as ^{57}Co is a very expensive radionuclide. In fact, he tried to recover the ^{57}Co from the urine of the patients.

V.R. McCREADY: Do you know how Dr. Nouel deals with the excreted urine from the point of view of radiation protection?

D. COMAR: I don't know.

K. HISADA: In Japan the ^{57}Co-bleomycin excreted into the urine has been recovered and re-used both for radiological and for economical reasons.

W.H. BEIERWALTES: It is not known exactly which fraction is routinely labelled in bleomycin, and part of the differences in results may be caused by the fact that each person is labelling a different fraction of bleomycin with a different specific activity. Native bleomycin is rich in copper, and we became interested in the copper content of the bleomycin which we used for labelling purposes. By neutron activation analysis we could show that the British preparation, which we used, was regularly free of copper. In our studies we also noticed that there was a late increase in tumour accumulation of the labelled bleomycin so we decided to use a long-lived copper nuclide, namely ^{67}Cu, the production of which, however, was very expensive. It is remarkable that a great number of papers at the meeting of the Society of Nuclear Medicine in 1973 dealt with bleomycin, but at the 1974 meeting I don't recall any paper at all on bleomycin studies. Finally, I would like to ask you, Dr. McCready, if you in your review have found any good studies on lung tumour to normal lung ratios? You of course know that bleomycin has a toxic effect on the lungs.

V.R. McCREADY: In lung tumours the results are about 70 to 90% correct, but the results are not all that better than the results obtained with ^{67}Ga. As to toxicity, you definitely get lung reactions with bleomycin and you can see them on plain X-rays from time to time. One of the interesting observations in Nouel's paper is that he had no uptake in cerebro-vascular accidents, whereas he had 44 positive results in a series of 45 glioblastomas.

H.J. GLENN: The In-bleomycin that we looked at originally and which was commercially available in the United States, had a three-day expiration period. Now it has been extended a bit, but still we are hampered by apparent poor stability of the compound. From reports it seems as if this is not so with the material used in England and other places.

K. HISADA: A report from Tokyo has shown the instability of In-bleomycin in vivo. The studies were based on blood samples several hours after injection of In-bleomycin.

H.J. GLENN: It is evident from several papers that the indium translocates from the bleomycin to transferrin within 6 hours in vivo.

V.R. McCREADY: As to the in-vitro stability, Cu, Ca and Fe (divalent and trivalent) have been added, and chromatographic analysis after 6 hours did not reveal any free indium.

E.H. BELCHER: I want to recall that the Harvard group working with $^{99}Tc^m$-tetracycline seems to be quite enthusiastic about this compound which they consider as the preferable radio-pharmaceutical for soft-tissue tumour detection in all regions except the abdomen. The mechanism for $^{99}Tc^m$-tetracycline tumour localization differs from that of ^{67}Ga-citrate, and it may be that the affinities of the two agents for various tumours differ. If this is the case, the combined use of the two agents may enhance the accuracy of tumour detection.

V.R. McCREADY: Perhaps a compound like $^{99}Tc^m$-tetracycline behaves differently from tetracycline itself and this may explain the good results. But the preparations may also vary from centre to centre. From what we read, we were not very impressed.

H.J. GLENN: In some studies carried out just before I left my institute, we found that tetracyclines are stable in solution at a pH of 1.5 to 2, whereas a partially colloidal solution may be formed at a pH between 4 and 7. This could explain the high liver uptake which is noted with tetracyclines at a neutral pH.

RADIOLANTHANIDES AS TUMOUR-LOCALIZING AGENTS

K. HISADA*, A. ANDO**, Y. SUZUKI*
Kanazawa University,
Kanazawa, Japan

Abstract

RADIOLANTHANIDES AS TUMOUR-LOCALIZING AGENTS.
In a systematic search for promising tumour-localizing agents, the tumour affinity of the higher-atomic-number radiolanthanides was discovered and the citrate form of ^{169}Yb has been used for clinical tumour scanning. The paper reviews the results of animal experiments, including other authors' reports, the clinical assessment of ^{169}Yb-citrate on 415 cases and ways of producing ^{167}Tm. The authors conclude that ^{167}Tm is a most promising tumour-scanning agent.

INTRODUCTION

A systematic search for a single radiopharmaceutical that would be incorporated into all kinds of malignant tumours, regardless of their anatomical location, has been made in our laboratory for several years. The elements tested so far in our laboratory are enclosed by a circle in the periodic table shown in Fig.1. The results have been summarized from the viewpoint of tumour affinity as follows: gallium and indium in Group III showed a very strong affinity for tumour, and scandium showed a slight affinity for Yoshida sarcoma; mercury and bismuth in Period 6 had a very strong affinity for tumour, and gold also had a very strong affinity for tumour when used as a $H^{198}AuCl_4$ solution [1]. Figure 2 shows our previous results in which elements in Period 6 show a very strong affinity for protein in in-vitro experiments. This accounts for the high tumour uptake of a heavy metal [1], but the affinity for the protein does not seem to play an important role in the uptake of the elements in Group III. Therefore, it is easily conceivable that the mechanism of tumour affinity of the elements of Group III differs from that of the elements of Period 6. In any case, on considering these facts carefully in relation to the periodic table, it seems logical to select thallium and the lanthanides as key elements. The ^{202}Tl-citrate did not show any affinity for Yoshida sarcoma, but some of the lanthanides, such as ^{170}Tm, ^{169}Yb and ^{177}Lu, showed excellent tumour affinity, true to our expectations.

Among 15 lanthanides, eight radioactive nuclides were available: ^{140}La, ^{141}Ce, ^{153}Sm, ^{153}Gd, ^{160}Tb, ^{170}Tm, ^{169}Yb and ^{177}Lu. Table I gives the results of this study [2]. Data for ^{67}Ga-citrate are shown in the last line for comparison. All the radionuclides show some affinity for the malignant tumour, and the retention value in the tumour tissue of ^{170}Tm-citrate is the highest being 1.34%/g. This is followed by ^{169}Yb-citrate and ^{177}Lu-citrate. Generally, the lanthanides with higher atomic numbers have a stronger affinity for a tumour than the ones with lower atomic numbers such as ^{140}La, ^{141}Ce and ^{153}Sm. The tumour-to-organ concentration ratio is also very high for ^{170}Tm-citrate and ^{169}Yb-citrate.

---

* Department of Nuclear Medicine, School of Medicine.
** Department of Radiological Technology, School of Paramedicine.

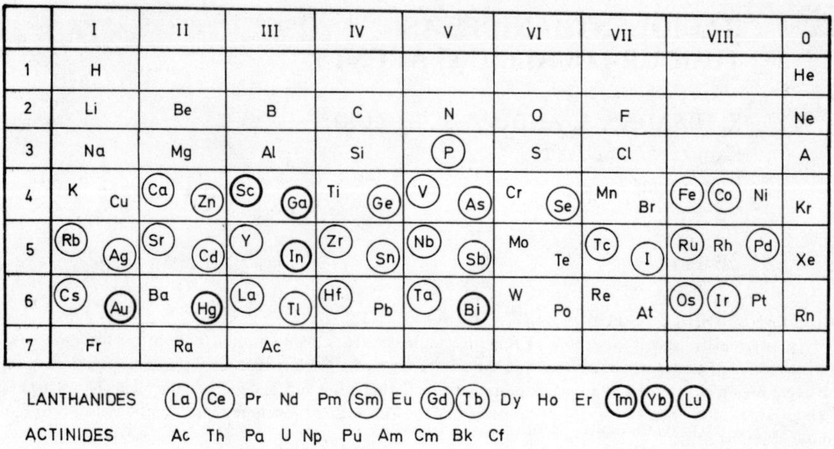

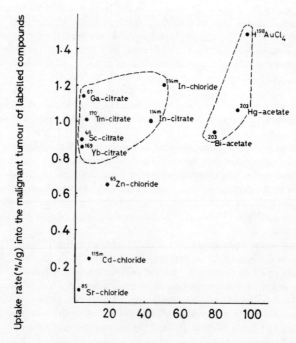

FIG.1. The elements in the periodic table that are enclosed by a circle have been screened for tumour affinity in our institute. The thick circles indicate that the element is tumour specific — note the position of these elements. Elements not enclosed will be investigated in the future.

FIG.2. Relationship between the uptake of a labelled compound in a malignant tumour and the affinity of metal compounds for protein.

TABLE I. DEPOSITION OF EACH LANTHANIDE AND ^{67}Ga IN YOSHIDA SARCOMA-BEARING RAT 24 HOURS AFTER INJECTION

(Modified from Ref. [2].)

	Specific activity	Retention values in tumour (%/g)	Ratio of tumour to blood	Ratio of tumour to muscle	Ratio of tumour to liver	Ratio of tumour to kidney	Ratio of tumour to bone
^{140}La-chloride	3 µCi/ La 40 µg	0.36	10.2	39.8	0.05	0.6	
^{141}Ce-citrate	2 µCi/ Ce 7.2 µg	0.38	23.1	15.1	0.04	0.3	0.7
^{153}Sm-citrate	3 µCi/ Sm 36 µg	0.44	19.8	10.5	0.05	0.2	0.6
^{153}Gd-citrate	2 µCi/ Gd 0.006 µg	0.37	20.4	18.2	0.17	0.3	0.1
^{160}Tb-citrate	2 µCi/ Tb 5.4 µg	0.41	19.9	7.4	0.29	0.2	0.1
^{170}Tm-citrate	33 µCi/ Tm 3 µg	1.34	94.7	54.6	2.5	1.9	0.4
^{169}Yb-citrate	2 µCi/ Yb 11 µg	0.72	52.2	38.7	1.9	0.9	0.4
^{177}Lu-citrate	30 µCi/ Lu 3 µg	0.59	26.4	23.3	1.3	0.9	
^{67}Ga-citrate	Carrier free	1.14	6.3	22.8	0.8	1.5	0.7

Following our report [2], Hayes et al. [3] and Yano et al. [4] presented their results which supported our observation that the higher-atomic-number radiolanthanides concentrate preferentially in tumour tissue. Hayes' group studied eight radiolanthanides and Fig. 3 is a histogram of the distribution of various radiolanthanides in liver, bone and 5123C hepatomas [5]. As early as 1956, Durbin at al. [6] found that the lighter lanthanides were taken up primarily in the reticuloendothelial system while the heavier lanthanides, such as Dy, Tm, Er, Lu, Yb and Tb, were taken up primarily in bone. The results of Hayes et al. [5] and our own results [2] confirm Durbin's observation: there was a dramatic decrease in the concentration of the radiolanthanides in liver as the atomic number of the nuclides increased, whereas the concentration in the tumour increased markedly, ^{170}Tm showing the highest tumour concentration. There were some small differences in detail between our results and those of Hayes et al. owing to the differences in tumour model.

To compare more precisely the distributions in vivo of ^{67}Ga and radiolanthanide ions such as ^{169}Yb or ^{170}Tm without any individual variation and tumour growth difference, we administered ^{67}Ga and ^{169}Yb or ^{170}Tm simultaneously as a mixture in the same chemical form as citrate to rats bearing Yoshida sarcoma. The specimens from the sacrificed rats were determined twice: once immediately after autopsy and once 12 days later. The amount of the two nuclides was quite easily determined by using the difference in half-lives of the two nuclides. Figure 4 shows that the retention of ^{169}Yb-citrate in the tumour tissue was similar to that of ^{67}Ga-citrate. A big

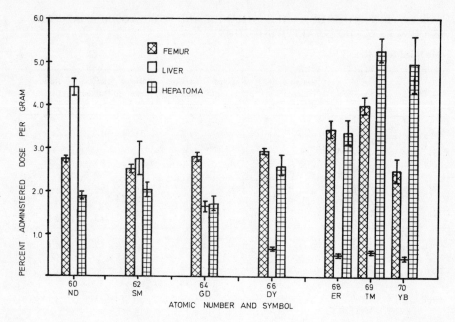

FIG.3. Concentrations (24 h) of high-specific-activity rare-earth radionuclides in femur, liver and tumour of male Buffalo rats bearing Morris 5123C hepatomas (groups of four animals). The concentrations were normalized to a body weight of 250 g. (From Ref. [5].)

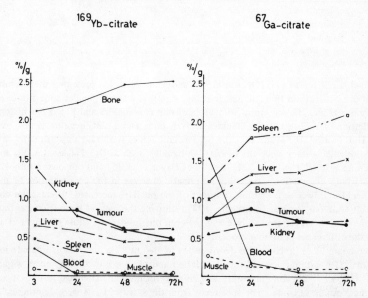

FIG.4. Results obtained after injecting ^{169}Yb-citrate and ^{67}Ga-citrate simultaneously into Yoshida sarcoma-bearing rats. The ordinate is the retention of radionuclide expressed as per cent of administered dose per gram of tissue in various tissues and a tumour. The abscissa is the time interval after administration of ^{169}Yb and ^{67}Ga. Note the marked differences in distribution of the two scanning agents in the normal tissues (From Ref. [2].)

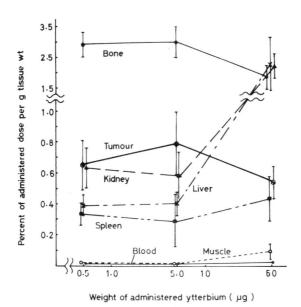

FIG.5. *Effect of ytterbium carrier on the distribution of ^{169}Yb in tissues of rats bearing Yoshida sarcoma. (From Ref. [7].)*

difference in distribution in normal tissues was noticed between ^{169}Yb and ^{67}Ga. The ^{169}Yb-citrate is cleared from the peripheral blood more rapidly than ^{67}Ga-citrate. Retention of ^{169}Yb-citrate in the liver and spleen was less than that of ^{67}Ga-citrate, whereas accumulation of ^{169}Yb in the bones was almost two times that of ^{67}Ga. These differences may cause the lower body-background in soft tissues and the denser bone image on the scintigram in the case of ^{169}Yb-citrate. ^{170}Tm-citrate also showed more or less the same results.

It would be interesting to know whether a carrier does have an effect on the distribution of lanthanides in-vivo as shown with ^{67}Ga-citrate. Figure 5 shows that the concentration of ^{169}Yb was greatly increased in liver and kidney and slightly decreased in tumour and bone at 50 µg per capita [7]. Hayes' group also found that there was a rather dramatic increase in the concentration of ^{171}Er in liver and spleen above ~ 10 µg Er/kg, but the tumour concentration was not affected. This carrier effect was also seen when stable erbium (at the microgram level) was administered with high-specific-activity ^{170}Tm [5]. Therefore, it will be important to use carrier-free or high-specific-activity preparations also in tumour scanning with radiolanthanides.

Another important point is whether or not lanthanides show selective accumulation in inflammatory tissue. An aseptic inflammation was induced by subcutaneous administration of 0.05 ml of croton oil. Table II reveals that ^{169}Yb-citrate and ^{170}Tm-citrate were accumulated to some degree in the inflammatory tissue but less than was ^{67}Ga-citrate. However, accumulation in the inflammatory tissue was something like one half or one quarter of that in the tumour.

Although the mechanism of the localization of radiolanthanides in the tumour is not fully understood, it would appear on the basis of the just-mentioned results that the higher-atomic-number radiolanthanides could be used for clinical tumour scanning. As promising nuclides, ^{157}Dy, ^{165}Er, ^{167}Tm, ^{175}Yb, ^{169}Yb and ^{177}Lu could be selected. For practical reasons, ^{169}Yb-citrate was selected for a clinical trial. ^{169}Yb is readily available and is a less expensive reactor-produced nuclide, the others are not commercially available.

TABLE II. DEPOSITION OF ^{67}Ga-, ^{169}Yb- AND ^{170}Tm-CITRATE IN RATS WITH INFLAMMATION AND WITH TUMOUR 24 HOURS AFTER INJECTION

(Unpublished data.)

	Aseptic inflammation (%/g)	Yoshida sarcoma (%/g)	Ratio of inflammation to tumour
^{67}Ga-citrate	0.56	1.17[a]	0.48
^{169}Yb-citrate	0.39	0.86[b]	0.45
^{167}Tm-citrate	0.20	0.78	0.26

[a] Average value of 42 animals.
[b] Average value of 39 animals.
The other figures are average values of 5 animals.

CLINICAL ASSESSMENT OF ^{169}Yb-CITRATE

Shortly after the discovery of the tumour affinity of the higher-atomic-number radiolanthanides, we tried to use ^{169}Yb-citrate for clinical tumour scanning [8]. Our success has promoted the distribution of ^{169}Yb-citrate to more than 20 hospitals in Japan.

The results of 360 confirmed cases of malignant tumour and 55 confirmed cases of benign lesions were collected from 10 different hospitals and assessed. As the results are given in more detail elsewhere [9], they are only outlined here. One hundred to 1000 µCi of ^{169}Yb-citrate were injected in the patients and the scintigrams were made by means of a rectilinear scanner or a gamma camera or both. Different scanning times after injection were adopted according to the protocol of each hospital. The scans were classified into 5 grades by a fixed standard as follows: (+++) means that the lesion was visualized more clearly than bone, (++) means that the lesion was visualized with the same density as bone, (+) means that the lesion was visualized clearly, but its activity was less than that of bone, (±) means that the activity of the lesion was not clearly seen, and (−) means that the lesion was not visualized.

The detection rate of the different types of tumour in different anatomical regions was calculated. In malignant tumours the overall positive rate was 65.3%, but the positive rate differed from hospital to hospital, from 33.3% to 89.5%. There was no correlation between the type of machine used for scanning and its positive rate, and no correlation between the injected dose of ^{169}Yb-citrate and the positive rate of the scanning.

In different anatomical regions, good results were obtained in the extremities (100%), pelvic area (100%), head and neck (78.5%) and lung (77.8%), but the positive rate in the abdomen was poor (48.3%). Among different histological types, the best positive rate was obtained in squamous cell carcinoma (77.5%), see Table III. The bone was visualized clearly in most of the scans in this series. A high level of bone activity sometimes made it difficult to detect an area of abnormal activity adjacent to bone, but the skeletal uptake was always useful for defining the anatomical orientation of the lesion.

In the 55 benign lesions the average false-positive rate was 29.1%; however, this figure varied from 0% to 62.5% from hospital to hospital. There was no significant correlation between the false-positive rate and the different scanning protocols such as the injected dose of ^{169}Yb-citrate, the type of machine and the scanning time after injection (Table IV).

TABLE III. THE POSITIVE RATE OF THE VARIOUS TYPES OF MALIGNANT TUMOURS IN DIFFERENT ANATOMICAL REGIONS

(From Ref. [9].)

	Squamous cell carcinoma		Adeno carcinoma		Undifferentiated carcinoma		Malignant lymphoma		Other malignant tumours		Number of cases	Positive rate (%)
	Pos.	Neg.	Pos.	Neg	Pos.	Neg.	Pos.	Neg.	Pos.	Neg.		
Head and neck Chest	6	1	1	3	1	1	3	1	0	0	14	78.5
Lung (primary)	20	4	4	0	4	2	0	0	0	0	37	75.5
Lung (secondary)	3	1	3	0	0	0	0	0	1	0	8	87.5
Mediastinum	2	1	1	0	0	0	0	0	1	1	6	66.7
Abdomen	3	4	13	16	0	0	0	2	11	11	60	48.3
Pelvic area	3	0	1	0	0	0	0	0	1	0	5	100
Extremities	1	0	3	0	0	0	0	0	1	0	5	100
Total	38	11	26	19	5	3	3	3	15	12	135	65.9
Positive rate	77.5%		55.3%		62.5%		50.0%		55.5%		65.9%	

TABLE IV. RESULTS OF THE TUMOUR SCANNING WITH ^{169}Yb-CITRATE IN BENIGN LESIONS COLLECTED FROM FIVE DIFFERENT HOSPITALS

(From Ref. [9].)

Hospitals	+++	++	+	±	−	No. of cases	Negative rate
Kanazawa Univ.	0	0	5	0	17	22	77.3%
Kurume Univ.	0	0	3	5	6	14	78.6%
Okayama Saiseikai	0	1	2	2	4	9	66.7%
Asahikawa Kosei	2	1	2	0	3	8	37.5%
Jikei Univ.	0	0	0	0	2	2	100.0%
Total	2	2	12	7	32	55	70.9%

False positive 29.1%

In our experience, quite a high false-positive rate (43.2%) is found in ^{67}Ga-citrate scanning (Table V). Consequently, ^{169}Yb-citrate was used in the hope of overcoming this problem, but without success. Table V also gives the false-positive rate in tumour scanning with ^{57}Co-bleomycin.

For the sake of brevity only one example of tumour scanning with ^{169}Yb-citrate is given in this paper. The case was a 46-year-old male with colon cancer. A barium enema revealed irregular narrowing in the middle of the ascending colon (Fig. 6A). An ^{169}Yb-citrate scan (250 μCi, 72 hours) of the lower abdomen demonstrated abnormal radionuclide localization in front of the right ilium

TABLE V. RECENT RESULTS OF TUMOUR SCANNING IN OUR DEPARTMENT
FROM MARCH 1973 TO MARCH 1974

(Unpublished data.)

Radiopharmaceutical	Positive	False negative	False positive[a]
^{67}Ga-citrate	112/123 (91.1%)	11/123	16/37 (43.2%)
^{169}Yb-citrate	47/73 (64.4%)	26/73	5/22 (22.7%)
^{57}Co-bleomycin	24/36 (66.7%)	12/36	3/8 (37.5%)

[a] "False positive" means benign lesions with accumulation of the radiopharmaceutical in the lesion.

(Fig. 6B). Visualization of the lumbar vertebrae and the sacroiliac joint was quite useful as anatomical landmarks in this case. The patient was operated 1 week after the scanning. The resected specimen was assayed in a well-type scintillation counter and the activity ratio of the normal colon to tumour was 1 to 6.

As a tumour-scanning agent, ^{169}Yb-citrate has several advantages: (1) ^{169}Yb is quite an economical nuclide because of its long shelf-life and its production by reactor, so ^{169}Yb can be available any time when tumour scanning is required. This may be very important in a clinical situation. (2) The main photon peak of ^{169}Yb is 198 keV which is suitable for scanning. (3) The soft-tissue background is extremely low and clear tumour images can be expected, especially 2–3 days after injection. On the other hand, the main disadvantage of using ^{169}Yb-citrate is a relatively high radiation dose to the patient. According to published data, the estimated absorbed whole-body dose is 1.74 rads/500 μCi and the skeletal dose 5.77 rads [8]. This whole-body dose is comparable to that obtained with ^{75}Se-selenomethionine. A positive rate in tumour scanning of 65.3% in malignant lesions may not completely satisfy the clinicians.

Considering the physical properties of ^{167}Tm and the distribution in vivo (Tables I and II), ^{167}Tm-citrate should be the nuclide of choice for clinical tumour scanning provided that it could become commercially available.

PRODUCTION OF THULIUM-167

In 1971, on the basis of animal studies with ^{170}Tm-chloride, Chandra et al. [10] proposed the use of ^{167}Tm for bone scanning, although ^{167}Tm was not commercially available at that time. In 1972, they succeeded in producing sufficient ^{167}Tm for medical use in a carrier-free form. By using enriched ^{167}Er$_2$O$_3$ and 15-MeV protons (p, n reaction), up to 75 μCi/μA·h could be produced in a small cyclotron [11]. In 1973, Steinberg et al. [12] reported the production of ^{167}Tm by two different nuclear reactions using a cyclotron: (1) α-particle bombardment of ^{165}Ho (α, 2n reaction), and (2) deuteron bombardment of enriched ^{167}Er (d, 2n reaction). The former ^{165}Ho (α, 2n) ^{167}Tm process was abandoned, but the reason was not given in the paper. The latter method was then used exclusively. They also succeeded in obtaining bone scans 24–28 hours after intravenous injection of 500 μCi of ^{167}Tm-citrate.

Based on a patient imaging dose of 500 μCi, the whole-body dose and skeletal dose were estimated to be 0.7 and 3.5 rads respectively [12].

For the purpose of soft-tissue tumour scanning we have endeavoured to produce ^{167}Tm with a linear accelerator at the Laboratory of Nuclear Science, Tohoku University, Sendai, Japan [13].

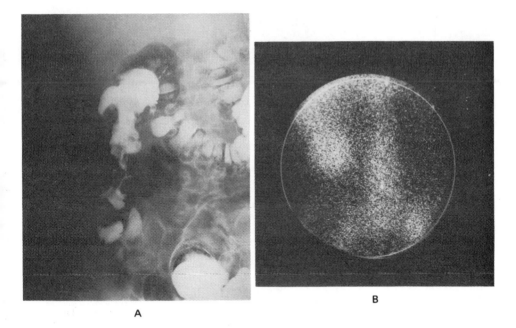

FIG.6. Colon cancer in a 46-year-old man. A. Irregular narrowing in the middle of the ascending colon revealed by a barium enema. B. ^{169}Yb-citrate scan of lower abdomen showing abnormal radionuclide localization in front of the right ilium.

The ^{168}Yb (γ, n) ^{167}Yb $\xrightarrow[17.7 \text{ min}]{EC}$ ^{167}Tm reaction using a linear accelerator has the advantage that it does not produce the long-lived ^{168}Tm. The ^{171}Yb (γ, p) ^{170}Tm reaction is also conceivable as a side reaction but its yield is presumably fairly low. Actually, we were able to obtain several hundred μCi of ^{167}Tm immediately after irradiation of γ-rays (60 MeV, 250 μA, 22 hours), though this amount was not enough for clinical purposes but only for an animal distribution study [13]. Some of the results of the animal experiments are shown in Table II. Further investigations on ^{167}Tm production will be reported in the near future. Yano and Chu [14] have also started to produce ^{167}Tm for tumour scanning by a ^{165}Ho $(\alpha, 2n)$ ^{167}Tm reaction at the LBL 88-in cyclotron. But the production yield of ^{167}Tm was only 15 -- 30 μCi/μA·h, which made the availability of ^{167}Tm difficult and expensive. They confirmed the potential usefulness of ^{167}Tm for tumour or bone scanning in animals. Quite recently, Chandra et al. [15] have reported on clinical bone scans in five patients with 300 – 500 μCi of ^{167}Tm-HEDTA.

CONCLUSION

The recent development of ^{99}Tc^m-polyphosphate, -pyrophosphate and -diphosphonate have made the usefulness of ^{167}Tm less important for bone scanning. Nevertheless, ^{167}Tm will remain a key radionuclide for soft-tissue tumour scanning and its mass production at a reasonable price is an urgent task in this field.

REFERENCES

[1] ANDO, A., HISADA, K., Distribution of gold and bismuth in rats subcutaneously transplanted with Yoshida sarcoma, and relations between the uptake ratio of various labeled compounds into the malignant tumor and their binding capacity to the protein, Radioisotopes **20** (1971) 321 (in Japanese).
[2] HISADA, K., ANDO, A., Radiolanthanides as promising tumor scanning agents, J. Nucl. Med. **14** (1973) 615.
[3] HAYES, R.L., et al., A comparison of the tissue distribution of ^{67}Ga and the rare earth radionuclides, J. Nucl. Med. **15** (1974) 501 (Abstract).
[4] YANO, Y., CHU, P., ANGER, H.O., Thulium-167: cyclotron production, chemical separation, and uptake in tumor mice, J. Nucl. Med. **15** (1974) 545 (Abstract).
[5] HAYES, R.L., et al., Radiopharmaceutical development – tumor-localizing agent, Oak Ridge Associated Universities 1973 Research Report ORAU-123, p.56.
[6] DURBIN, P.W., et al., Metabolism of the lanthanons in the rat, Proc. Soc. Exp. Biol. Med. **91** (1956) 78.
[7] ANDO, A., HISADA, K., Affinity of lanthanons for malignant tumor (II), Radioisotopes **21** (1972) 684. (in Japanese).
[8] HISADA, K., et al., Tumor scanning with ^{169}Yb citrate, J. Nucl. Med. **15** (1974) 210.
[9] HISADA, K., SUZUKI, Y., et al., A clinical evaluation of tumor scanning with ^{169}Yb citrate, Radiology **116** (1975) 389.
[10] CHANDRA, R., et al., ^{167}Tm: a new bone scanning agent, Radiology **100** (1971) 687.
[11] CHANDRA, R., et al., Production of ^{167}Tm for medical use, Int. J. Appl. Radiat. Isot. **23** (1972) 553.
[12] STEINBERG, M., et al., "^{167}Tm-citrate for bone imaging", Radiopharmaceuticals and Labelled Compounds (Proc. Symp. Copenhagen, 1973) **2**, IAEA, Vienna (1973) 151.
[13] SAKAMOTO, H., et al., Production of High Specific Activity ^{167}Tm and its Affinity for Tumor and Bone, Research Report of Laboratory of Nuclear Science, Tohoku University, 7 1 (1974) (in Japanese).
[14] YANO, Y., CHU, P., Cyclotron produced thulium-167 for bone and tumor scanning, Submitted to Int. J. Appl. Radiat. Isot.
[15] CHANDRA, R., et al., Evaluation of ^{167}Tm-HEDTA as a bone scanning agent in humans and its comparison with ^{18}F, Br. J. Radiol. **47** (1974) 51.

DISCUSSION

E.H. BELCHER: A paper by Sullivan et al. has just been published from Argonne National Laboratory in the January 1975 issue of the International Journal of Nuclear Medicine and Biology. The authors have studied the distribution of lanthanide and actinide radioisotopes in tumour-bearing mice. The results indicate a remarkable differentiation between even such chemically similar species as the lanthanide compounds. The behaviour of samarium citrate in sarcoma-bearing mice and in sarcoma-bearing rats is strikingly similar. The four-valent state is much more specific than the five- or six-valent states. The heavier lanthanides are more tumour-specific than the light lanthanides. The radioisotopes of samarium, ytterbium and neptunium localize much more in tumours than gallium and should show lower body-backgrounds and be better scanning agents than ^{67}Ga. The authors feel that the lanthanide and actinide compounds show promise as tumour diagnostic agents for clinical use.

W.H. BEIERWALTES: It is extremely important that the same specific activities are used when you compare results from one radionuclide with those of another one. The uptake obtained (per cent of dose/gram) depends very much on the specific activity. Every synthesis may have a different specific activity, and results may vary if we don't have the same specific activity as in previous studies. The ^{170}Tm specific activity in Dr. Hisada's studies was more than ten times greater than that of the other compounds (with the exception of ^{177}Lu), and it also showed the highest retention in the tumour (per cent of dose/gram). We have had the same experience with labelled testosterone where we got a seven-fold increase in the uptake in the dog prostate when we used a ten-fold higher specific activity. The same is true in adrenal scanning with ^{125}I-iodo-cholesterol in the dog where the activity concentrations rose from 0.05 μCi/g to 3.75 μCi/g, when we changed the specific activity. Finally, I guess the reason why you have used most of your time on ^{169}Yb-citrate is because ^{172}Tm was not available to you for routine studies as early as you had the ytterbium preparation for your experiments.

H.J. GLENN: From the standpoint of pharmaceutical units, I should like to suggest that we speak about "assay" when we talk about μCi accumulation in the tissues and that we speak about "concentration" when we speak about mass units. We should not indiscriminately use the word concentration for activity as well as for mass.

R.L. HAYES: Dr. Hisada, you have presented a very fine paper. I noticed you didn't consider the possibility of using ^{171}Er. There are various shortcomings to this radionuclide, but we consider this nuclide to be worth looking at clinically. It is true that it is a β-emitter, but the half-life is only 7.5 hours, and the γ-energy is quite adequate for rectilinear scanning. One of the advantages of this radionuclide would be that it could be administered in multi-mCi quantities, for instance 5 – 10 mCi. I too should like to add that Dr. Edwards has looked at ^{157}Dy in a few patients. Most of these patients had also had ^{67}Ga scans. Unfortunately, the ^{67}Ga scans appeared superior to the ^{157}Dy scans. In these studies ^{157}Dy was given as a citrate.

W.H. BEIERWALTES: I should like to ask you, Dr. Hayes, if you have tried ^{170}Tm at Oak Ridge in a comparison with ^{67}Ga.

R.L. HAYES: ^{170}Tm is impractical to use in humans because of its long half-life and low energy. But in animal experiments ^{170}Tm and ^{67}Ga compare quite favourably. ^{67}Ga had an absolute tumour concentration in our animal models, about twice that of ^{170}Tm. But the tumour-to-non-tumour ratios were superior for ^{170}Tm since ^{170}Tm clearance from normal tissues is much more rapid than is that of ^{67}Ga.

W.H. BEIERWALTES: Is ^{157}Dy made from the reactor or from the cyclotron?

R.L. HAYES: It can be made either way. The cyclotron route is obviously more expensive. The ^{157}Dy that we have used has been reactor-produced from separated ^{156}Dy. This initially posed a special problem since we could not obtain separated ^{156}Dy in sufficient purity until recently when Oak Ridge National Laboratories devised a new mass separator technique. The new technique got rid of the problem of ^{164}Dy contamination. Incidentally, this new technique will make available practically any isotope in a very high state of purity, even if the isotopic abundance in the natural element is very low.

H.J. GLENN: I would like to ask you, Dr. Hisada, if you have any information on how the Yb-citrate was synthesized. When we tried in our laboratory to make Yb-citrate exactly the way we made Ga-citrate, just by taking Yb-chloride in acid solution, adding citrate to it, and neutralizing it, we found that almost all of it ended up in the liver as a colloid. Then, in another laboratory, we designed another means of preparing Yb-citrate that didn't end up in the liver. When this was achieved, we found that we had ^{169}Yb accumulation in bone about twice the concentration of ^{99}Tc^m-pyrophosphate. The results showed an uptake of about 50 – 60% of the dose in bones rather than 30% that was achieved with the standard Tc-pyrophosphate. This bone localization was probably the result of carrier – compare the use of low specific activity material. In addition, I would like to recall some work by Subramanian who used cyclotron-produced ^{157}Dy but changed the biological distribution by adding Lu as the carrier and thus made it a marrow-scanning agent.

K. HISADA: I don't know exactly how the Yb-citrate was made but I think my associates dried a ^{169}Yb-DTPA solution in a flask and then added conc. HNO$_3$. The conc. HNO$_3$ was evaporated by heating and a solution of sodium citrate was added to form ^{169}Yb-citrate. The Yb-citrate is now commercially available in Japan. It has a pH of 4 to 5 and contains 2 μg/ml of citric acid.

H.J. GLENN: We, too, decreased the pH from 7 to about 4, and then added about triple the amount of citrate to keep the Yb in solution. This once more shows that you have to be extremely careful in your formulations for bio-distribution studies.

TUMOUR LOCALIZATION USING COMPOUNDS LABELLED WITH CYCLOTRON-PRODUCED SHORT-LIVED RADIONUCLIDES

D. COMAR
CEA, Département de biologie,
Service hospitalier Frédéric Joliot,
Orsay, France

Abstract

TUMOUR LOCALIZATION USING COMPOUNDS LABELLED WITH CYCLOTRON-PRODUCED SHORT-LIVED RADIONUCLIDES.
  The properties of an ideal tumour-localizing agent are summarized with special reference to the use of ^{18}F-, ^{15}O-, ^{11}O- and ^{13}N-labelled compounds. The increased number of "medical" cyclotrons will allow extensive research in this field. Representative examples are given of the labelling techniques and of the clinical usefulness of compounds labelled with cyclotron-produced short-lived radionuclides.

INTRODUCTION

  The localization of tumours by radioactive agents implies access to a molecule which is labelled with a radionuclide of suitable half-life and gamma energy and which, at a given moment, has a stronger affinity for diseased tissue than for the surrounding normal tissue.
  The more specific the agent and the stronger and faster its fixation to tumours, the greater is its value. In fact, early diagnosis of a tumour is only possible if enough radioactive substance is fixed per unit mass to allow detection of the diseased tissue as soon as it is formed. At present there do not seem to be any radiopharmaceutical products that fulfil these criteria. The reason for this may be that the radionuclides generally used are not normal components of biological molecules and are therefore not incorporated specifically in the cell metabolism: their fixation in a tumour only reflects a greater vascular permeability or an increased cell metabolism, phenomena which often fail to demonstrate the difference between a temporary inflammation and a malignant formation. The appearance of compact cyclotrons for medical use in hospitals has opened up new possibilities, since these are the only instruments capable of producing gamma-emitting radionuclides of the main components of living matter, namely ^{15}O, ^{11}C and ^{13}N, often associated with ^{18}F.
  An estimate of the potentialities of these four radionuclides for tumour detection is based on a detailed analysis of their characteristics, the chemical and physical properties of their stable isotopes and the technical and economic problems of their production and use.

1. SOME ASPECTS OF THE PRODUCTION AND PROPERTIES OF THE RADIONUCLIDES ^{15}O, ^{13}N, ^{11}C AND ^{18}F

  The position of ^{11}C, ^{13}N and ^{15}O in isotope charts and particularly their immediate proximity to a natural element of large isotopic abundance possessing one proton less, suggest that their production yield by charged-particle reaction of the (d, n) type will be good (Fig. 1). Fluorine-18

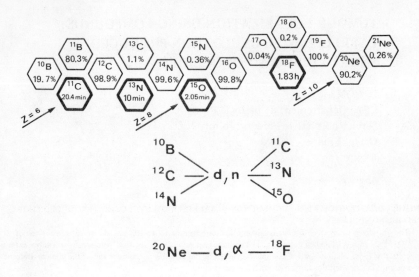

FIG.1. Nuclear reactions leading to the formation of ^{11}C, ^{13}N, ^{15}O and ^{18}F by deuteron bombardment.

cannot be produced in any great quantity by the same reaction because of the scarcity of the isotope ^{17}O, but large amounts can be obtained by the (d, α) reaction on ^{20}Ne. The deuteron energy necessary for these reactions to take place must be higher than the coulomb barrier of the target elements, which is less than 3 MeV for the deuteron or proton irradiation of elements of Z below 10. Since, in addition, these reactions are energy-liberating, or slightly energy-absorbing, the incident deuteron energy may be limited and hence a low-power machine can be used. In fact, in certain cases, for practical reasons and for ease of handling, it may be an advantage to use the (p, α) reaction on nitrogen to produce ^{11}C and the (^{3}He, n) or (^{4}He, 2n) reactions on oxygen to prepare ^{18}F.

These remarks show that the short-lived nuclides ^{15}O, ^{13}N, ^{11}C and ^{18}F can be manufactured without difficulty for the needs of diagnostic medicine by means of a particle source accelerating deuterons to 8 or 10 MeV maximum or, as the need arises, deuterons to 10 MeV and protons to 15 MeV. The strength of the current necessary to produce enough radioactivity is given in Table I which shows the yield in $\mu Ci/\mu A \cdot min$ for the nuclides ^{11}C, ^{13}N and ^{18}F obtained routinely in our laboratory.

A second property of the short-lived nuclides ^{11}C, ^{13}N, ^{15}O and ^{18}F arises directly from the manner of their preparation and from their half-lives. Being obtained by nuclear transmutation, their specific radioactivity in the physical sense of the term is equal to 1. This concept has in fact no chemical or biological meaning, the only important criterion being that of the mass of element which accompanies the radioactivity.

Since the mass of a radioelement per unit of radioactivity is directly related to its physical half-life it is obvious that the shorter the half-life the smaller the mass accompanying this radioactivity. This is clearly shown in Table II which gives the radioactivities corresponding to one micromole of several radionuclides obtained by nuclear transmutation.

Actually, for the nuclides ^{11}C, ^{15}O and ^{13}N, which correspond to the most abundant elements in the biosphere, the source of contamination by stable elements is such that it is impossible to obtain the theoretical activity per unit mass. In fact, it is only necessary for this radioactivity per unit mass to be higher than that measurable by the most sensitive analytical methods available.

TABLE I. PRODUCTION YIELD OF ^{11}C, ^{13}N AND ^{18}F

Nuclide	Nuclear reaction	μCi/μA·min
^{11}C	^{14}N(p, α)^{11}C[a]	2000
^{13}N	^{12}C(d, n)^{13}N[b]	230
^{18}F	^{20}Ne(d, α)^{18}F[c]	200

[a] $N_2 + 5\% \; O_2$; P = 1 bar; flow rate: 1 litre/min; target l = 70 cm.
[b] CO_2; P = 2 bars; target l = 40 cm.
[c] Ne; P = 3 bars; quartz target l = 70 cm.

TABLE II. SPECIFIC ACTIVITY IN Ci/μM OF A FEW CARRIER-FREE RADIONUCLIDES

Radionuclide	Half-life	Ci/μM
Carbon-11	20.4 min	9.2×10^3
Nitrogen-13	10 min	1.9×10^4
Oxygen-15	2.05 min	9.2×10^4
Fluorine-18	1.83 h	1.7×10^3
Carbon-14	5.730 years	6.2×10^{-5}
Tritium	12.26 years	2.9×10^{-2}

One of the advantages of these very high specific activities is that short-lived radionuclides can be used to label molecules which cannot be employed at present because of their high toxicity or their very low natural concentration in living organisms. This property seems to be very important for the study of tumour localization, several authors having shown the fundamental influence of the specific radioactivity of a labelled molecule on its fixation. Another feature common to these four radionuclides is their mode of disintegration. They are positron emitters and, on disintegration, emit two 510-keV gamma rays at 180° which can be measured in coincidence, the localization and resolution advantages of this being well known. In addition, as practically one positron is emitted per disintegration the gamma radiation at equal radioactivity is double that of a radioelement disintegrating by β^- and γ emission. For the same number of photons emitted, the radioactivity necessary in ^{15}O, ^{13}N, ^{11}C and ^{18}F will be at most half that of a radioelement with no β^+ emission. This fact can be favourable from a dosimetric point of view. However, owing to the relatively high energy of the β^+ particles and the radiation emitted by these four radionuclides, the dose absorbed by the tissues is not negligible in spite of their short half-lives. This is illustrated in Table III which shows the β^+ and γ doses absorbed by the whole body and the liver for 1 mCi of radioelement fixed.

TABLE III. ABSORBED DOSE IN rad/mCi OF ^{11}C AND ^{18}F HOMOGENEOUSLY DISTRIBUTED IN WHOLE BODY AND LIVER

Radioelement	Dose absorbed by:	
	Whole body	Liver
^{11}C	0.011 { β^+ 0.0055 ; γ 0.0055 }	0.34 { β^+ 0.24 ; γ 0.10 }
^{18}F	0.033 { β^+ 0.014 ; γ 0.019 }	1.35 { β^+ 0.81 ; γ 0.54 }

Not enough work has been done yet on tumour localization with short-lived radionuclides to allow a general approach. However, it can be foreseen that the compounds to be used should fulfil the following three criteria:

The labelling time must not exceed two or three half-lives of the radioelement. The choice of labelling position is thus restricted to easily accessible chemical functions.

The biological phenomenon to be detected must be observable in less than 3 to 4 half-lives of the labelled product. This means that the blood clearance of the substance must be fast and that fixation in the tumour must be strong.

The result anticipated must supply new or better quality information than that obtained with radio-elements or labelled molecules now available.

2. EXAMPLES OF PRESENT AND POTENTIAL APPLICATIONS OF SHORT-LIVED RADIO-NUCLIDES TO THE LOCALIZATION OF TUMOURS

(a) Simple chemical combinations of short-lived radioelements

Fluorine-18. The value of ^{18}F in the form of NaF for skeleton visualization and the discovery of bone tumours has been known for a long time, and methods for its preparation can be found in various journals [1]. Sodium fluoride is still the most widely used of the ^{18}F-labelled pharmaceutical products for preparing images of the skeleton and detecting bone metastases. Compared with various existing products such as ^{99}Tc^m-labelled pyrophosphate, diphosphonate and polyphosphate, ^{18}F has certain advantages. Since it is produced without carrier it produces no pharmacological effects, unlike the ^{99}Tc^m-phosphates injected at weighable doses which can have a cardiovascular action in for instance old people. Whereas the images obtained with ^{18}F are reproducible, the ^{99}Tc^m-polyphosphate distribution in the skeleton has been found to vary from batch to batch without it being known whether this is due to a variation in the specific activity or to a difference in the length of the poly-phosphate chain. Finally, the accumulation of polyphosphates in the kidneys makes the images more difficult to interpret than those obtained with ^{18}F [2]. ^{99}Tc^m-labelled derivatives are actually much more widely used at present because of their low cost and ease of distribution, though it is probable that at equal availability and with no detection problems for 510-keV γ-emitters, ^{18}F would be considered a better diagnostic agent. During the last three years several comparative studies on the detection of tumours and bone metastases with ^{18}F and other radiopharmaceuticals (^{99}Tc^m-Sn-EHDP, ^{167}Tm-HEDTA) have not always proved the superiority of ^{18}F fluoride, although the advantages of this tracer have been shown elsewhere [3, 4]. The increased osteoblastic activity of bone metastases leads to a larger deposit of calcium and phosphorus in the bone. The use of all these radiopharmaceuticals, including ^{18}F, aims at the detection and possibly quantification of this phenomenon, but it seems that all research in this direction has been terminated since the introduction of ^{99}Tc^m-labelled complex phosphates as indicators [5].

Oxygen-15. The very short half-life of this radionuclide (2.05 min) prevents both the synthesis of complex molecules and its use for the exploration of slow phenomena.

However, dynamic information can be obtained from a static observation by constant administration of a short-lived radiopharmaceutical. The advantage of using equilibrium images (those obtained several half-lives after the start of constant administration) is that high-quality static images of dynamic processes can be obtained [6]. This idea has also been investigated by Russ et al. [7] who placed dogs in isotopic equilibrium by letting them take in air labelled with ^{15}O or C^{15}O. The results show that images obtained using ^{15}O in the stationary state depend on the metabolism and perfusion of the organ considered. This method would thus be potentially useful for the detection of hyperactive tissues or fast cell division processes.

(b) **Complex chemical combinations of short-lived radioelements**

The incorporation of ^{11}C and ^{13}N in natural compounds or drugs is at present an obstacle in the development of methods for tumour localization. Conventional organic chemistry is not very suitable for ultra-rapid syntheses, and chemists are concentrating more on the improvement of standard synthesis methods than on the search for new molecules with an affinity for tumours. The main ^{11}C-labelled precursors now used in the synthesis of complex molecules are ^{11}C-formaldehyde and methyl iodide for the preparation of ^{11}CH$_3$ groups, ^{11}C-cyanhydric acid for the production of amino acids, nitriles (Strecker's synthesis) and aliphatic primary amines with labelled carbon preceding the amino group. Finally, ^{11}CO$_2$ can be used to create the carboxyl group on aliphatic chains, aromatic nuclei and hetero-rings. The interesting molecules to label would be those that are incorporated rapidly into the increased metabolism of the cancerous cell, a category including the amino acids and nucleosides. Thus, methyl-thymidine has been labelled in the CH$_3$ group [8]. ^{11}C-aspartic acid has also been recommended for tumour detection but the fast incorporation of ^{11}C into an amino acid seems to be difficult. This is probably why an attempt was made to incorporate ^{13}N into glutamic acid, alanine and glutamine by enzymatic synthesis from the NH$_4^+$ ion. The synthesis of ^{13}N-ammonia and the incorporation of radioactive nitrogen in the organic molecule are so fast that several tens of mCi of ^{13}N-ammonium chloride and glutamine may be obtained. These substances have revealed tumours like adenocarcinomas and lymphosarcomas in dogs [9], and ^{13}N-glutamic acid has demonstrated bone tumours in the same animals [10]. ^{13}N-alanine is believed to be adequately fixed by the pancreas and could be used to detect tumours of this organ. ^{11}C-labelled dopamine [8], which is known to have an affinity for the adrenal-medulla, and sulphoanilide analogues [11] could be used for the localization of phaeochromocytomas and neuroblastomas.

Fluorine-18-labelled molecules are especially interesting for the detection of tumours, the half-life of ^{18}F being appreciably longer than that of the other radioelements discussed (T = 1.83 hour).

Chemically, fluorine is the most electronegative of the halogens and therefore polarizes its bond with other atoms. When incorporated in drugs or hormones it often causes an increase in reactivity and more pronounced pharmacological actions.

Compared with iodine, a very widely used tracer in nuclear medicine, fluorine has a high carbon bonding energy (107 to 121 kcal/mole), whereas that of the C–I bond is only 57 kcal/mole. A fluorinated molecule is thus more stable than one containing iodine.

The radius of the fluorine atom is 1.36 Å, much smaller than that of iodine (2.16 Å), and its mass is about 7 times lower. Thus it has less effect on the structure and is less bulky.

The chief difficulty encountered is the incorporation of carrier-free fluorine in an organic skeleton. However, a certain number of fluorinated amino acid analogues acting as antimetabolites have been synthesized by exchange with the stable fluorinated compound (fluorophenylalanine, fluorotryptophane, fluorotyrosine, fluoro-DOPA). Some have been used without success to visualize the pancreas, while no medical results have been published on the others.

As with ^{11}C, puric and pyrimidne derivatives have been labelled with ^{18}F. 5-Fluoro-uracil, fluorocytosine and fluoroadenine [12] have been synthesized in this way.

Attempts have been made to label certain sex steroids with ^{18}F [13] but no results have been published so far.

CONCLUSION

^{15}O, ^{11}C, ^{13}N and ^{18}F, being for a long time considered as "exotic" nuclides, have only been studied in laboratories close to the particle accelerators generally used by physicists.

Within the next two years some twenty compact cyclotrons, intended exclusively for the manufacture of radionuclides for medical use, will be installed throughout the world in or near hospitals and will thus give access to very short-lived radioelements.

Already a number of arguments and proofs exist to show the superiority of these nuclides over radioelements that are more easily available and hence more commonly used. Moreover, it is essential for the medical physicist to know that, in spite of the very short half-lives of these radionuclides, it is possible to incorporate them into natural compounds or drugs without altering the molecular structure and to administer them in doses so small that they do not give any detectable metabolic perturbations or toxic symptoms.

REFERENCES

[1] SILVESTER, D.J., "Accelerator production of medically useful radionuclides", Radiopharmaceuticals and Labelled Compounds (Proc. Symp. Copenhagen, 1973) 1, IAEA, Vienna (1973) 197.

[2] QUINN, J.L., CREWS, M.C., WESTERMAN, B.R., "^{99}Tc^{m} versus ^{18}F bone-scanning agents: advantages and disadvantages", Radiopharmaceuticals and Labelled Compounds (Proc. Symp. Copenhagen, 1973) 1, IAEA, Vienna (1973) 119.

[3] McNEIL, M., CASSADY, J.R., GEISER, C.F., JAFFE, N., TRAGGIS, D., TREVES, S., Radiology 109 (1973) 627.

[4] MONTE BLAU, GANATRA, R., BENDER, M.A., Semin. Nucl. Med. 2 1 (1972) 31.

[5] ESTEBAN-VELASCO, J., "Positive diagnosis of tumours: a historical review", Proc. 1st World Congress of Nuclear Medicine, Tokyo, 1974, World Federation of Nuclear Medicine and Biology, p. 115.

[6] JONES, T., BROWNELL, G.L., TER-POGOSSIAN, M.M., J. Nucl. Med. 15 6 (1974) 505 (Abstract).

[7] RUSS, G.A., BIGLER, R.E., TILBURY, R. S., McDONALD, J.M., LAUGHLIN, J.S., "Whole-body scanning and organ imaging with oxygen-15 at the steady-state", Proc. 1st World Congress of Nuclear Medicine, Tokyo 1974, World Federation of Nuclear Medicine and Biology, p. 904.

[8] WOLF, A.P., CHRISTMAN, D.R., FOWLER, J.S., LAMBRECHT, R.M., "Synthesis of radiopharmaceuticals and labelled compounds using short-lived isotopes", Radiopharmaceuticals and Labelled Compounds (Proc. Symp. Copenhagen, 1973) 1, IAEA, Vienna (1973) 345.

[9] CHRISTIE, T.R., MONAHAN, W.G., GELBARD, A.S., CLARKE, L.P., LAUGHLIN, J.S., J. Nucl. Med. 15 6 (1974) 483 (Abstract).

[10] McDONALD, J.M., CLARKE, L.P., CHRISTIE, T.R., GELBARD, A.S., LAUGHLIN, J.S., J. Nucl. Med. 15 6 (1974) 515 (Abstract).

[11] ICE, R.D., WIELAND, D.M., BEIERWALTES, W.H., LAWTON, R.G., J. Nucl. Med. 15 6 (1974) 503 (Abstract).

[12] SILVESTER, D.J., CLARK, J.C., PALMER, A.J., "The future of accelerator-produced radiopharmaceuticals", Proc. 1st World Congress of Nuclear Medicine, Tokyo, 1974, World Federation of Nuclear Medicine and Biology, p. 181.

[13] SHONE, L.L., WINCHELL, H.S., J. Nucl. Med. 7 (1966) 336 (Abstract).

DISCUSSION

V.R. McCREADY: When your speak about ^{13}N-alanine for tumour localization, do you look for a negative or a positive area in the case of pancreas?

D. COMAR: A positive area.

V.R. McCREADY: This could not be specific to the pancreas.

D. COMAR: It is known that alanine is taken up to a higher concentration by pancreas than by most other tissues, and perhaps more rapidly. The incorporation of alanine in pancreas tumours might be higher than in normal pancreas tissue.

T. MUNKNER: Dealing with these very short-lived nuclides we once more touch upon the importance of a regional high clearance. You will never get more into the tumour (if you inject intravenously) than that fraction which is equal to the ratio between local blood flow and the cardiac output, maybe even less due to an extraction below 1.0. You can, of course, increase the accumulation by local intra-arterial injection or (theoretically) by increasing the local fraction of the total cardiac output.

D. COMAR: As a matter of fact we have never tried to detect tumours with short-lived radionuclides. We have, however, labelled drugs with ^{11}C and looked at the distribution in the body. Some of these drugs were studied because they have an affinity for brain tissues; in these cases the blood clearance was very high. Two minutes after the injection the level of activity in the blood had fallen by a factor of 100 or perhaps 1000, as a large amount of the labelled compounds had been taken up by the kidneys or the liver. In these cases we had a very high concentration in the brain, and the brain-to-blood ratio was, maybe, 100:1 three to five minutes after the injection. I don't know whether we can extrapolate, but we may hope to find a similar difference in tumours.

T. MUNKNER: Even if you have a high brain-to-blood ratio, I think that you have only a tiny fraction of the injected dose in the brain.

D. COMAR: In some cases it has been 7 to 10% of the injected dose.

T. MUNKNER: This figure is pretty close to the brain blood flow fraction of the total cardiac output. This means that you have an extraction close to 100%.

D. COMAR: The uptake may increase if we inject locally and if we can diminish the binding of the labelled compound to proteins. It is known for example for drugs that they must be unbound in order to penetrate the blood brain barriers if you want to have a very quick uptake.

T. MUNKNER: Are there any means of creating a local dilatation of the vessels?

V.R. McCREADY: Tumour vessels don't react to pharmaceuticals in the same way as normal vessels do.

E.H. BELCHER: Is there anyone who could comment on the costs in relation to cyclotron-produced short-lived radionuclides?

W.H. BEIERWALTES: One principal problem in physics and instrumentation versus radiopharmaceuticals is that man seems to enjoy and develop instruments far more fancifully and effectively than he does the type of thinking that has to go into the development of a specific metabolic process of a tumour. We have π meson therapy units which cost millions and millions of dollars with so far almost no detectable results, and many people have developed beautiful radionuclide-labelled agents that have made a dramatic improvement in diagnosis, such as ^{99}Tc^m-polyphosphate, with research grants totalling US$ 30 000 per year. A cyclotron which opened up recently has cost US$ 650 000, and the personnel to run it contributes by an additional permanent expenditure of US$ 65 000 a year. Most of us would be delighted to have a US$ 65 000 grant a year to develop radiopharmaceuticals.

D. COMAR: I think we have to consider two different kinds of cyclotrons. On the one hand you have the very short-lived radionuclides that are cyclotron-produced: oxygen, carbon, fluorine and nitrogen. For these four isotopes you don't need very large machines. You need just a deuteron machine or a proton machine with a fixed energy, that is a small thing with not too high a current.

This kind of machine does not exist for the moment, but I think if we can prove in a few years that some important things can be done with radionuclides produced by cyclotrons, this machine will exist at a low price. On the other hand, what we have for the moment in different hospitals are very complex machines which are very expensive and which are able to make quite a number of radionuclides, also those with long half-lives. Long half-life radionuclides need not be made at a hospital, you can just as well buy them from a commercial firm. The small machines which I mentioned need not be very expensive and need not have a very sophisticated personnel. We haven't been able to create these small machines yet, but I hope they will come.

If you change from using ^{131}I to using ^{123}I, you will not discover new things because a lot has already been done with the other radioisotopes of iodine. It is different with short-lived radionuclides because new vistas are opened up due to the very high specific activity and due to the fact that you are able to label drugs and able to label biological molecules without changing their metabolism. It is an area of new features and possibilities. It is indeed quite different from comparing ^{123}I with ^{131}I or from replacing Ga by In.

R.L. HAYES: Usefulness as well as availability and cost will be keys to this new area. The unique aspects of the short-lived radionuclides of which we speak are that they constitute structural elements in organic and biological compounds. In addition, they are generally positron emitters. I hope that we are on the verge of a break-through in instrumentation for positron detection. A combination of availability of short-lived radionuclides with positron emission and improved instrumentation for the detection of positrons would be highly beneficial.

H.J. GLENN: It is our experience that one may get strange results when one labels with elements that are not physiological. This is a point which counts in favour of true physiological studies with ^{14}C and with the short-lived radionuclides of carbon, nitrogen and oxygen.

T. MUNKNER: There are additional areas where you could use short-lived radionuclides, for instance in respiratory studies, without being forced to incorporate the short-lived nuclides into specific compounds. The cyclotrons which were used for studies in theoretical physics are (at least in part) no longer of any use for studies in physics, and they could possibly be transformed into cyclotrons for medical purposes.

W.H. BEIERWALTES: We do have two cyclotrons close to our institute, one is small and the other one is a huge research cyclotron. All support for the huge cyclotron and the additional laboratories have been cut off. And we have, in addition, the problems of a rapid radiochemistry, which is necessary if you want to have the compounds with short-lived radionuclides at your disposal within, say, 5 min.

D. COMAR: The dopamine analogues, which you have used, might be labelled in 10 min with ^{11}C, instead of labelling them with iodine.

W.H. BEIERWALTES: We have, in fact, been working in this area for about six years showing that dopamine is accumulated in the adrenal medulla of the dog, in neoblastoma in diagnostic and therapeutic concentrations, and in phaeochromocytomas. We have had a full-time organic chemist working on this project but we still have no way of quickly incorporating ^{11}C into those compounds which are now of primary interest to us.

D. COMAR: I think the labelling with ^{11}C of some of these compounds can be done in about 20 min. Some of your compounds have in their side-chains a dimethylamino group, and it can easily and rapidly be labelled with ^{11}C, maybe in 10 min.

W.H. BEIERWALTES: The trainee who worked in our department with the labelling of dopamine went to another centre later and since then we haven't heard about this work.

D. COMAR: Such work calls for chemical and medical people. There are some places where the medical people are willing to use these labelled compounds, and other places where the chemical people are able to make the compounds, but only few places where both categories are present at the same time.

H.J. GLENN: In my institution we have looked at the possibility of dismantling a cyclotron which was no longer used for physical studies, of shipping it a considerable distance to our institute, and then reassembling it and get it in working order. The total cost of this transfer amounted to more than the amount required for building a new medical cyclotron. We have also tried to find a cyclotron that would satisfy the needs of our high-energy physics people, our therapy people and our radioisotope production needs. At the present time there is no cyclotron available that would satisfy all these fields.

SUMMARY OF GENERAL DISCUSSION

The future development of tumour-localizing agents was discussed in several sessions on the basis of the papers presented and the Group's background of personal experience with presently available techniques. Special attention was paid to the role the International Atomic Energy Agancy might play in promoting the development of new agents, in supporting research in individual fields, and in making the techniques available on a world-wide scale.

1. FUTURE STUDIES

The importance of displaying primary and secondary tumours is evident, and the role of nuclear medicine techniques for this purpose was emphasized. However, it was stressed that guidelines are needed for prospective studies on tumour-localizing agents to obtain the full benefit of clinical trials without wasting too much effort during the next ten years. Future clinical trials might for instance be supervised by national societies of nuclear medicine or be organized regionally in order to prevent uncoordinated efficacy trials as has occurred in the past. In setting up guidelines to evaluate tumour-localizing agents, much can be learned from the recommendations laid down in the ^{67}Ga intercomparison study in the United States, the lung scanning intercomparison study, which went on for five years, and from the radiological health thyrotoxicosis follow-up study, which went on for twelve years. Trials are expensive, and some very firm data bases are needed for comparison and proper evaluation of studies with new agents or techniques. More insight into cancer demography and cancer biology will be necessary for prospective studies. It may be advisable to split the research tasks between a number of competent departments or laboratories in such a way that some types of cancer are preferentially studied in one or more of these centres, for instance with concentrated research on endocrine tumours, anaplastic tumours and on tumour biology in different departments. There will, too, be a need for more extensive studies on animals with specific animal tumours which have been shown to bear the best relationship to human tumours, in contrast to the present studies which for the greater part have been carried out on patients. The guidelines could be established as a protocol, where experts in bio-statistics could help to define how many cases are needed to answer specific questions. Unfortunately, most of the present clinical trials have been set up without a preliminary statistical evaluation and the statistician has been called only when it was time to publish the results and then, of course, it was too late.

2. RADIOPHARMACEUTICALS

In order to make it possible to obtain a real comparison of results, there should be one common source of the radiopharmaceutical which should be subjected to careful quality control. This would assure that the same kind of material is given to all the patients. The compound that is shipped by a firm is not necessarily the same compound that the investigator uses when he injects it. If the compound is left under adverse conditions, it may deteriorate and the results could be entirely different from those in another laboratory. If in one institution the technicians do the injections and in another institution the investigator takes the same responsibility, it is nearly certain that varied amounts of the compound will be left at the injection site. It will be important to specify the use of carrier and to have standard methods for reporting the results on tumour uptake as tumour-to-non-tumour ratios with reference to a basic standard, for instance data based on the use of ^{67}Ga or perhaps radioiodinated serum albumin. From past experience it is known how difficult a clinical evaluation of radiopharmaceuticals may be. Even with carefully worked out protocols and patient reports, and with the full consent of all investigators, some of the case reports sent to an evaluation committee may be completely impossible to interpret because some of the investigators don't want to perform the study in any other way than their own.

3. ANIMAL TUMOUR MODEL SYSTEMS

Animal tumour model systems using radionuclides should be studied more intensively in an effort to relate animal tumour models to human tumours to help separate more clinically predictive tumour models from models of doubtful effectiveness. The better tumour models should be more extensively studied, standardized and described from the standpoint of origin, biochemistry, biophysiology, immunology and mechanism of tumour uptake, and more complete descriptions of tumours be included in all publications. A better understanding of tumour function in relation to radioactive substances and the use of standardized, well-characterized clinically predictive tumours would hopefully lead to the design of more specific radiopharmaceuticals for tumour-localization studies. There is also a need for standardization in studies using animal tumour models with regard to research techniques and protocols, methods of reporting results of tumour and tissue uptakes, the use of "internal" standards such as ^{131}I-iodinated serum albumin, and the use of an internal tissue ratio such as the ratio of tumour-to-liver content, in all tumour uptake studies for various animal tumour models. These needs for standardization could perhaps best be met by the formation of a permanent international panel to study and suggest guidelines for future research using radionuclides and animal tumour models.

4. PATIENT MATERIAL

In setting up a protocol for the studies, it is rather easy to define some of the parameters, such as specific activity and dose per kilogram of body weight. One of the next necessary steps will be to define the patient material. Even in common diseases, where the nomenclature is fairly well worked out, there is a lack of uniformity in almost every class of tumour. Comparisons should be made with patients in the same stage of disease, and final tabulations should be based on well-defined criteria for what constitutes a "positive" and a "negative" scan. Histological proof of the disease is important, but this is unrealistic in many clinical situations. The interval between the scan procedure and the other procedures with which the scintigraphic results are compared is important and not always recorded. It is not necessary in a clinical trial that every site of the tumour and metastases be confirmed – indeed there is no means for doing this. As to the interpretation of the results, the guidelines should be as precise as possible and should pay special attention to the range of normal, which is indeed often wide, the appreciation of which takes time and experience. This too, has to do with an unsolved problem in nuclear medicine, i.e. that perception may change in given persons from time to time. Some groups exert considerable energy and aggressiveness in demonstrating a site that they strongly suspect is present, whereas other groups are not so extensive in their efforts. Professional reward is another matter of concern. A team can submerge an individual who for obvious reasons may be concerned with certain rewards and prestige as well as status and advancement in his career, and these ambitions may be thwarted by team publications and by the fact that he is only one "out of ten" authors and perhaps not the first one.

5. INSTRUMENTATION AND TUMOUR-LOCALIZING AGENTS

It was the consensus of the participants that new and important means for localizing tumours could be expected by the development of new radiopharmaceuticals rather than of instrumentation which, for the moment, seems to be at a plateau where the physical possibilities have been explored to a high degree of sophistication. It is important that the medical problem should be more strictly defined in order to tell the physicists where to improve upon the present scintigraphic equipment. Special emphasis should be paid to a greater sensitivity and a better spatial resolution for future equipment. Blood background subtraction methods have been used to increase the detectability of tumours, but have not always produced better results because of the inherent statistical problems. It would be desirable to have something definite said about the preference for either ^{123}I or ^{131}I, and this would have a significant impact on the suppliers of radiopharmaceuticals and on nuclear medicine in general. Also, it would be worth while exploring the possibilities for using positron emitters. The specific activity of the radiolabelled compound has a profound influence on the final scintigraphic results, and the best results seem to be obtained with compounds with a high specific activity. It is an observation of great importance that a small change in a compound may dramatically change its per cent uptake in a tumour. By careful studies of such changes, tumour uptake might be improved by minor alterations in the chemical structure of agents. It is interesting, too, to study the lipid-to-water partition coefficient as those agents that are least water-soluble lend themselves best to tumour uptake, perhaps because the cell membrane is, in part, lipoprotein in nature. In the development of new radiopharmaceuticals more attention should be paid to the use of radionuclide-labelled antibodies against specific tumour antigens as well as to the unique property of metastatic or primary bone tumours to cause osteoblastic activity, which makes it possible to make a scintigraphic display of bone tumours. Another unique property of tumour tissue may be related to the action of irreversible enzyme inhibitors which, in some cases, allow a many-fold increase in the concentration of certain radionuclides in a specific location. In the search for a new labelled compound it is natural to look at metabolic pathways in general. A systematic approach to further development of the present drugs, as done by the drug companies, may also be promising. The companies take the first compound that gives promising results and begin in a systematic way to synthesize all related compounds, changing one portion of the structure at a time and then sending the different compounds away for pharmacological evaluation. In a number of cases this has brought a final success, for instance in the development of radiolabelled compounds which enter the adrenal medulla and also have a pharmacological effect. It is reasonable to consider the pharmacological effect when searching for a diagnostic tumour-localizing agent. Usually the conclusion will be wrong, but one may also stumble upon some compounds which distribute in the right way. Another approach might be to look into the textbooks and try all the compounds that have been discarded because of severe toxic effects. Radionuclide scintigraphy has the tremendous advantage of using tracer amounts of a compound. A very toxic drug can be used in nanogram amounts and, when labelled, display proper localization without toxic effects. More studies should be carried out on specific plasma protein labels, inter alia with gallium. Reference was also made to studies on the use of labelled macrophages, even though the present results show that this involves a difficult in-vitro procedure. A final problem with radiopharmaceuticals is that of formulation. After the first successful adrenal scan in a dog it took about two years to formulate properly the iodocholesterol compound for human use.

6. ADMINISTRATION MODE

The possibilities for improving the scintigraphic results by modifying the way in which the labelled compound is administered were discussed in some detail. Normally, the highest regional uptake is obtained when the radiolabelled compound is given inter-arterially in the vessels leading to the organ, a somewhat lower uptake is achieved when the compounds are given intravenously, and the lowest uptake is found after oral administration. Most of the compounds have to be given intravenously, at least for practical reasons. Repeated injection or constant infusion would presumably not improve the results significantly, and it would make the procedures much more cumbersome. The greatest success would certainly be achieved by designing an agent which binds in a unique way, or in as unique a way as possible, to the tumour as compared to normal tissues, and then to study how the compound could be cleared from all the other tissues at a relatively uniform rate. This statement was exemplified by the experience around 1960 showing that successful images of dog kidneys could be obtained by continuous infusion of ^{131}I-Hippuran, but that similar or better results were obtained by a single injection of ^{203}Hg-chloromerodrin. After 1960 no one went on with continuous infusion for kidney scintigraphy, and everyone preferred to use the compound which concentrated selectively in the organ.

7. COMBINED TECHNIQUES

Simultaneous or successive applications of two or more labelled compounds with different mechanisms for the tumour uptake were also discussed. As an example, it was mentioned that bleomycin concentrates in perfused tumour and could be used in combination with labelled tetracycline, which concentrates in necrotic areas. But none of the participants had real experience in this area. It has been a success in chemotherapy to use several agents, and the principles for this approach in therapy could also be applicable to radionuclide studies. Studies along this line have been carried out in Mexico and other places for the differential diagnosis of liver lesions. There are several points which distinguish the use of two agents in chemotherapy from a similar approach in nuclear medicine. In chemotherapy the goal is to block alternative metabolic pathways. This principle could in fact be used in scintigraphy by for instance blocking one enzymatic pathway for the production of an adrenal steroid and waiting for the pituitary response to the deficiency by an increased ACTH stimulation which would then enhance another synthetic pathway of the steroids in the adrenal through a different enzymatic activity. This approach might result in a higher concentration of a labelled precursor in the adrenal gland. This principle might be used with even greater success if irreversible enzyme inhibitors could be used at the same time. In a similar way, radionuclide therapy might then have a tremendous advantage over chemotherapy, if it were possible to give an irreversible enzyme inhibitor that concentrates diagnostically. When this compound is used in pharmacological doses, the organs may develop an alternative pathway. But, if a therapeutic radionuclide is attached to the irreversible enzyme inhibitor, the radiation will devitalize the neoplastic cell before the cell has had a chance to develop an alternative pathway. By this means the cells may be successfully knocked out before they can develop an alternative pathway, whereas in chemotherapy this is extremely difficult as the neoplastic cell opens up an alternative pathway as soon as the primary pathway is blocked.

8. PERFUSION AND CELLULAR UPTAKE

Part of the discussion was devoted to ideas on improving the uptake by changes in regional perfusion or in the local conditions for diffusion of the labelled compounds into the tumour cells. Decreased blood flow may in some cases increase the local extraction. In tumours that are well perfused, a change of the permeability would no doubt enhance the tumour uptake. In tumours that are poorly perfused it will be difficult to improve the situation as tumour vessels don't react to normal stimuli. More studies are needed on the influence of perfusion on "hot" and on "cold" tumours as opposed to the uptake in the surrounding normal tissues. The possibilities for changing the binding of a labelled compound to plasma proteins will have a significant influence on diffusion. It is known from early studies that gallium is bound to plasma proteins. To reduce this binding and promote deposition in bone, it was reasonable to try to block with stable gallium or to find a suitable substitute which was less toxic. After trials with various ions which have the same oxidation state, the same ionic radius, etc., scandium was found to be the most effective agent for blocking the binding sites for gallium on plasma proteins. When gallium was found to concentrate in soft-tumour tissues, additional experiments were carried out to study the effect of scandium on tumour uptake. In this case scandium did not decrease the binding of ^{67}Ga in the tumour, and it was natural to go on with animal experiments in which ^{67}Ga injection was supplemented by injections of stable scandium in an amount of about 0.5 mg/kg. This could be done without toxic effects. By doing so, the clearance of gallium from the blood increased to such an extent that it even became possible to consider using ^{68}Ga, with its very short half-life, to visualize tumours. In man, however, it was shown that scandium, at the very low concentration (about 100 μg/kg) at which it begins to be effective in clearing the blood, has also a haemolytic effect on human erythrocytes. This haemolytic effect was never found in animal studies. Even if scandium was not sucessful in human studies in increasing the clearance of gallium from blood, the principle of having a simultaneous administration of a blocking agent may some time in the future be of great value in increasing the tumour-to-non-tumour ratio. It must, however, be remembered that if the clearance is increased, for instance by the use of competitive binding agents, the uptake in the cells may decrease if this is just due to diffusion. It is well known from some brain-tumour scanning models that a strong inverse relation exists between renal clearance and tumour uptake, the faster the clearance from the blood the less the uptake in the tumour. This of course favours compounds that are actively transported into the cell, then changed in a chemical way which permits them to stay intracellularly. The influence of renal clearance on the results also forms the basis for a suggestion about studying the influence of dehydration on uptake in the tumour.

9. AGENCY SUPPORT

The International Atomic Energy Agency can promote the use of radionuclide-labelled compounds for tumour localization in several ways. A comprehensive list of departments or institutes of nuclear medicine might facilitate mutual exchange of ideas and results and also form the basis for distribution of publications of general interest. It is often important to know about research and studies carried out in other laboratories and to be able to find where special experience can be obtained in other departments. It was stressed that the International Atomic Energy Agency doesn't work in such a way that it can direct research in the field of tumour-localizing agents, but rather by supporting studies and research carried out in existing institutes. The Agency's primary obligation is to support studies in developing countries. One of the ways to promote future activities in tumour localization would be to have meetings of working groups with a frequency depending on the development of new radiopharmaceuticals or of new equipment. These working groups might clarify the present state of the art and give ideas about the most rewarding lines of future research. It was felt that the IAEA symposia had contributed essentially to dissemination of the results obtained and that the symposia in the coming years should be devoted more to the tasks and to the radiopharmaceuticals than to instruments which have already been described in sufficient detail at earlier meetings. In addition, the Agency should try to establish guidelines on how to report on the results. This kind of information would facilitate the intercomparisons between results obtained in a number of departments. Finally, the Agency was asked to consider possibilities for facilitating world-wide use of good radiopharmaceuticals by supporting production of relevant products and by making them available to all countries at a reasonable cost, and, if possible, try to improve upon the logistics of providing labelled compounds. However, past experience has shown that the provision of labelled compounds, even if simple in principle, is often difficult in practice, and the obstacles in providing a world-wide supply have added to the restraints for the use of radionuclide-labelled compounds for tumour localization.

LIST OF PARTICIPANTS

W.H. BEIERWALTES
Section of Nuclear Medicine,
University of Michigan Medical Center,
Ann Arbor, MI 48015,
United States of America

D. COMAR
CEA, Département de Biologie,
Service Hospitalier Frédéric Joliot,
91406 Orsay, France

H.J. GLENN
Section of Nuclear Medicine,
The University of Texas System Cancer Center,
M.D. Anderson Hospital and Tumor Institute,
Houston, TX 77025,
United States of America

R.L. HAYES
Medical Division,
Oak Ridge Associated Universities,
P.O. Box 117,
Oak Ridge, TN 37830,
United States of America

K. HISADA
Department of Nuclear Medicine,
School of Medicine,
Kanazawa University,
Takara-machi 13–1,
Kanazawa, Japan

H. LANGHAMMER
Nuklearmedizinische Klinik und Poliklinik
rechts der Isar der Technischen
Universität München,
Ismaninger Strasse 22,
8000 München 80,
Federal Republic of Germany

V.R. McCREADY
Department of Nuclear Medicine,
The Royal Marsden Hospital,
Downs Road,
Sutton, Surrey
United Kingdom

REPRESENTATIVE OF WHO

M. SENTICI
WHO Liaison Officer with IAEA

CONSULTANT TO IAEA

T. MUNKNER
Department of Nuclear Medicine,
Rigshopsitalet 2033,
Blegdamsvej 9,
2100 Copenhagen,
Denmark

REPRESENTATIVES OF IAEA

R.M. KNISELEY Division of Life Sciences,
IAEA

E.H. BELCHER Division of Life Sciences,
(Scientific Secretary) Medical Applications Section,
IAEA

HOW TO ORDER IAEA PUBLICATIONS

An exclusive sales agent for IAEA publications, to whom all orders and inquiries should be addressed, has been appointed in the following country:

UNITED STATES OF AMERICA   UNIPUB, P.O. Box 433, Murray Hill Station, New York, N.Y. 10016

In the following countries IAEA publications may be purchased from the sales agents or booksellers listed or through your major local booksellers. Payment can be made in local currency or with UNESCO coupons.

ARGENTINA	Comisión Nacional de Energía Atómica, Avenida del Libertador 8250, Buenos Aires
AUSTRALIA	Hunter Publications, 58 A Gipps Street, Collingwood, Victoria 3066
BELGIUM	Service du Courrier de l'UNESCO, 112, Rue du Trône, B-1050 Brussels
CANADA	Information Canada, 171 Slater Street, Ottawa, Ont. K1A 0S9
C.S.S.R.	S.N.T.L., Spálená 51, CS-110 00 Prague Alfa, Publishers, Hurbanovo námestie 6, CS-800 00 Bratislava
FRANCE	Office International de Documentation et Librairie, 48, rue Gay-Lussac, F-75005 Paris
HUNGARY	Kultura, Hungarian Trading Company for Books and Newspapers, P.O. Box 149, H-1011 Budapest 62
INDIA	Oxford Book and Stationery Comp., 17, Park Street, Calcutta 16; Oxford Book and Stationery Comp., Scindia House, New Delhi-110001
ISRAEL	Heiliger and Co., 3, Nathan Strauss Str., Jerusalem
ITALY	Libreria Scientifica, Dott. de Biasio Lucio "aeiou", Via Meravigli 16, I-20123 Milan
JAPAN	Maruzen Company, Ltd., P.O.Box 5050, 100-31 Tokyo International
NETHERLANDS	Marinus Nijhoff N.V., Lange Voorhout 9-11, P.O. Box 269, The Hague
PAKISTAN	Mirza Book Agency, 65, The Mall, P.O.Box 729, Lahore-3
POLAND	Ars Polona, Centrala Handlu Zagranicznego, Krakowskie Przedmiescie 7, Warsaw
ROMANIA	Cartimex, 3-5 13 Decembrie Street, P.O.Box 134-135, Bucarest
SOUTH AFRICA	Van Schaik's Bookstore, P.O.Box 724, Pretoria Universitas Books (Pty) Ltd., P.O.Box 1557, Pretoria
SPAIN	Diaz de Santos, Lagasca 95, Madrid-6 Calle Francisco Navacerrada, 8, Madrid-28
SWEDEN	C.E. Fritzes Kungl. Hovbokhandel, Fredsgatan 2, S-103 07 Stockholm
UNITED KINGDOM	Her Majesty's Stationery Office, P.O. Box 569, London SE1 9NH
U.S.S.R.	Mezhdunarodnaya Kniga, Smolenskaya-Sennaya 32-34, Moscow G-200
YUGOSLAVIA	Jugoslovenska Knjiga, Terazije 27, YU-11000 Belgrade

Orders from countries where sales agents have not yet been appointed and requests for information should be addressed directly to:

Division of Publications
International Atomic Energy Agency
Kärntner Ring 11, P.O.Box 590, A-1011 Vienna, Austria